US CLIMATE CHANGE POLICY

Transforming Environmental Politics and Policy

Series Editors:

Timothy Doyle
Keele University, UK, and University of Adelaide, Australia

Philip Catney
Keele University, UK

The theory and practice of environmental politics and policy are rapidly emerging as key areas of intense concern in the first, third and industrializing worlds. People of diverse nationalities, religions and cultures wrestle daily with environment and development issues central to human and non-human survival on the planet Earth. Air, Water, Earth, Fire. These central elements mix together in so many ways, spinning off new constellations of issues, ideas and actions, gathering under a multitude of banners: energy security, food sovereignty, climate change, genetic modification, environmental justice and sustainability, population growth, water quality and access, air pollution, mal-distribution and over-consumption of scarce resources, the rights of the non-human, the welfare of future citizens—the list goes on.

What is much needed in green debates is for theoretical discussions to be rooted in policy outcomes and service delivery. So, while still engaging in the theoretical realm, this series also seeks to provide a 'real world' policymaking dimension. Politics and policymaking is interpreted widely here to include the territories, discourses, instruments and domains of political parties, non-governmental organizations, protest movements, corporations, international regimes, and transnational networks.

From the local to the global—and back again—this series explores environmental politics and policy within countries and cultures, researching the ways in which green issues cross North-South and East-West divides. The 'Transforming Environmental Politics and Policy' series exposes the exciting ways in which environmental politics and policy can transform political relationships, in all their forms.

Other titles in the series:

Energy, Governance and Security in Thailand and Myanmar (Burma)
A Critical Approach to Environmental Politics in the South
Adam Simpson

Energy Security in Japan
Challenges After Fukushima
Vlado Vivoda

US Climate Change Policy

CHRISTOPHER J. BAILEY
Keele University, UK

Routledge
Taylor & Francis Group

LONDON AND NEW YORK

First published 2015 by Ashgate Publishing

2 Park Square, Milton Park, Abingdon, Oxfordshire OX14 4RN
52 Vanderbilt Avenue, New York, NY 10017

Routledge is an imprint of the Taylor & Francis Group, an informa business

First issued in paperback 2020

British Library Cataloguing in Publication Data
A catalogue record for this book is available from the British Library.

The Library of Congress has cataloged the printed edition as follows:
Bailey, Christopher J.
  US climate change policy / by Christopher J. Bailey.
     pages cm. – (Transforming environmental politics and policy)
  Includes bibliographical references and index.
  ISBN 978-1-4094-4017-8 (hardback)
1. Climatic changes – Political aspects – United States. 2. Environmental policy – United States. 3. United Nations Framework Convention on Climate Change (1992 May 9). Protocols, etc. (1997 December 11) I. Title. II. Title: U.S. climate change policy.
  QC903.2.U6B35 2015
  363.738'745610973–dc23

                                                                    2015019285

ISBN 978-1-4094-4017-8 (hbk)
ISBN 978-0-367-59760-3 (pbk)

Contents

List of Figures and Tables

Figures

Tables

Series Editors' Preface

The beginnings of this series emerged at Keele University in a collaboration between Tim Doyle and Phil Catney. Since the late 1970s, Keele has been renowned across the globe as one of the leading universities engaged in teaching and research into the politics and international relations of the environment.

Our initial conversations with Ashgate were around two objectives. First, we wanted to transform the rather narrow, dominant conceptions of environmental politics and policy—particularly in the global North—by opening it up to include issues more central to traditional human politics and policymaking. Gone are the days when the 'environment' is something that people engage in (and with) as some kind of 'luxury' pursuit, when and if they have the time and the resources to do it. Nowadays, environmental issues—both in the North and the South—are front and centre. Secondly, we strongly felt that much needed in environmental debates was for theoretical discussions to be rooted in policy outcomes and service delivery. In short, the series would look at the exciting ways in which environmental politics and policy can transform governance, in all its forms.

Timothy Doyle, Keele University, UK, and University of Adelaide, Australia, and Philip Catney, Keele University, UK

Acknowledgements

Thanks are due to a number of people who helped with this project. I benefitted greatly from discussions with a number of my colleagues at Keele University about policymaking and climate change. Jon Herbert, John Vogler, Robert Ladrech, and Phil Catney helped me clarify my thoughts at different times. Phil Catney also read large parts of the manuscript and offered great advice. I would also like to thank Keele University for giving me the research leave that enabled me to complete the project. Finally, my wife Helen helped improve the figures and tables, and offered love and support.

List of Abbreviations

ACES Act	American Clean Energy and Security Act
AR	Arkansas
ARRA	American Recovery and Reinvestment Act
AZ	Arizona
CA	California
CAFE	Corporate Average Fuel Economy
CFCs	chlorofluorocarbons
CN	Connecticut
CNN	Cable Network News
CO	Colorado
CO_2	carbon dioxide
COP	Conference of the Parties
D	Democrat
DE	Delaware
DOT	Department of Transport
EO	executive order
EPA	Environmental Protection Agency
EU	European Union
FY	fiscal year
GA	Georgia
GCC	Global Climate Coalition
GDP	Gross Domestic Product
I	Independent
ID	Idaho
IL	Illinois
IN	Indiana
INC	Intergovernmental Negotiating Committee
IPCC	Intergovernmental Panel on Climate Change
KS	Kansas
LA	Louisiana
MA	Massachusetts
MD	Maryland
ME	Maine
MI	Michigan
MN	Minnesota
MO	Missouri
MT	Montana

NAS	National Academy of Sciences
NASA	National Aeronautics and Space Administration
NC	North Carolina
ND	North Dakota
NE	Nebraska
NES	National Energy Strategy
NJ	New Jersey
NM	New Mexico
NOAA	National Oceanic and Atmospheric Administration
NSPS	New Source Performance Standards
NV	Nevada
OH	Ohio
OK	Oklahoma
OMB	Office of Management and Budget
PA	Pennsylvania
PACs	political action committees
PSD	Prevention of Serious Deterioration
R	Republican
SAR	Second Assessment Report
SUVs	sports utility vehicles
TN	Tennessee
TX	Texas
UAW	United Automobile Workers
UN	United Nations
UNEP	United Nations Environmental Program
UNFCCC	United Nations Framework Convention on Climate Change
UT	Utah
VA	Virginia
VA-HUD	Veterans Affairs - Housing and Urban Development
VT	Vermont
WA	Washington State
WCP	World Climate Program
WMO	World Meteorological Organisation
WV	West Virginia
WY	Wyoming

Introduction

The United States government has frequently been accused of doing little to address concerns that greenhouse gas emissions are contributing to global warming. "Until recently, US climate change policy has appalled the world" is the opening sentence of one study (Driesen, 2010, 1). Another study claims that the United States "has been particularly unresponsive with respect to climate change" (Jamieson, 2011, 47). Prominent on the charge sheet presented by critics is the decision of the Bush Administration to withdraw from the Kyoto Protocol in 2001, the failure to enact legislation regulating the emission of greenhouse gases, and the apparent lack of a coherent national climate change policy. President Obama's inability to persuade the US Senate to pass a flagship proposal to establish a cap-and-trade system to regulate greenhouse gases in 2010 seemed to confirm this view of the United States government as lacking a commitment to tackle climate change. Partisan, institutional, economic, and cultural factors have all been advanced as explanations for this unwillingness to confront climate change in a meaningful manner. This narrative of inaction, however, does not tell the whole story. Although there is much to commend in the view of the United States government as a laggard when it comes to climate change, such a characterisation fails to tell the whole story. Over the last thirty years or so the United States government has spent billions of dollars on climate change research, measures to conserve energy, and developing renewable energy technology. Action has also been taken to limit greenhouse gas emissions from automobiles and power plants. The United States government has not consistently embraced the need to address climate change, has taken action in a piecemeal way accompanied by much kicking and screaming, and in the view of many environmentalists and scientists could do much more, but claims of inaction are difficult to sustain when the evidence is examined carefully.

In this book I trace and explain the development of the US government's actions to address climate change from the late 1970s onwards. My focus is on the making of policy. I am interested in why particular approaches to dealing with climate change have been chosen at particular times rather than evaluating the success of these choices. Parts of this path have been trodden by others. Full-length academic studies of climate change policy in the United States include Gerrard (2008), Rahm (2010), Sussman and Daynes (2013), and Brewer (2014), while Cass (2006) and Carlarne (2010) provide comparisons with European policy approaches. These works offer a considerable amount of material on the content of policy, and identify some of the factors at play in the politics of climate change. Two important elements, however, are missing from most of these studies. The first is an explicit recognition that different parts of the US government have tried to address climate change, and that the task therefore is to identify and explain *policies* rather than a *policy*. Sussman and Daynes (2013) and Brewer (2014) are exceptions as they examine the roles played by various domestic political actors in making US climate change policy. The second is a conceptual framework that ties together the various factors that contribute to the timing and shape of these policies. In this book I seek to redress both of these shortcomings. I identify and explain the actions on climate change initiated by various policy venues within the US government, and I employ Kingdon's (2011) "streams approach" to explain levels of government attention to climate change and the rise and fall of policy alternatives. Kingdon's conceptual framework has the benefit of showing how various factors interact over time to produce opportunities for policy making. Policy outputs depend upon the content and mix of these streams when "policy windows" open.

A number of broad insights about US government, politics, and culture flow from studying the development of climate change policies. Climate change poses a number of scientific, economic, and political challenges that test the capacity of the political system to respond in a meaningful way. Senator David Durenberger (R. MN) colourfully described dealing with climate change as "as easy as nailing jello to the wall" in a Senate Committee hearing in 1986 (SCEPW, 1986). Often described as a "wicked," or even "super wicked" problem that defies easy resolution because of uncertainty, conflict, and the fact that time is not costless when searching for solutions, climate change raises a host of difficult questions about the nature of science, faith in technology, American dependence on fossil fuels, the continued quest for economic growth, and obligations to future generations (Rittel and Webber, 1973; Lazarus, 2009; Levin et al., 2012). The search for answers to questions such as these reveals a great deal about the power of vested interests, partisan and ideological calculations, and cultural and religious views on environmental stewardship.

A brief discussion of terminology is important at this stage. In this book I employ the term climate change to describe the trend of increasing global temperatures attributed to human activity. No political purpose is intended by this choice of terminology. The usage is widely accepted and conforms to current US government practice. I have made no attempt, however, to change the language of participants in the policy process. Scientific papers, government reports, political speeches, and media stories have variously referred to the "greenhouse effect" and "global warming" in the past (and present), and I have stuck with the original usage when citing or quoting from these sources.

Identifying US Climate Change Policies

What is climate change policy? This question is overlooked in much of the literature on the subject but is fundamental to the task of tracing and explaining policy development. Without an answer that defines the subject in a clear way it is impossible to be confident about what constitutes evidence of climate change policy, where this evidence can be found, and when to start looking for it. Identifying climate change policy, however, is fraught with problems. Some of these problems are common to the study of public policy in general. Disputes about the nature of public policy abound and include arguments over the definition of policy and who is involved in the policy process (Cairney, 2012; Knill and Tosun, 2012). No simple solution is available that resolves these disputes and any definition of policy will raise a number of questions. For the purpose of this book I define policy broadly as action taken by public actors to address a particular issue. This can involve forcing organisations or individuals to take particular forms of action to solve a problem as well as efforts to persuade or "nudge" these organisations or individuals to take voluntary action (Thaler and Sunstein, 2008). Other problems are specific to climate change. Climate change is a multi-dimensional problem that is impossible to place within a neat analytical box, and this complicates the search for policy. In Dryzek's terminology it is "unbounded" (1990, 59). Action to address climate change may be located within proposals dealing with energy, transport, forestry, urban planning, and other policy areas as well as contained in something labelled as climate change or global warming.

The obvious starting point to the task of identifying climate change policy is to answer the question "What is climate change?" Disputes over the answer to this question have raged for decades with a number of sceptics dismissing the very idea that the climate is changing, some arguing about the cause of any changes, and others challenging the extent and consequences of predicted changes (Booker, 2010; Oreskes and Conway, 2010; Oreskes, 2011). The idea that the world's climate could change began to coalesce in the early nineteenth century. French physicist Jean-Baptiste Joseph Fourier identified the phenomenon that would later be termed "the greenhouse effect" in the 1820s, and Swiss naturalist Jean Louis Rodolphe Agassiz posited the idea of an Ice Age a decade later (Fleming, 1998; Hulme, 2009). Subsequent work by the Irish scientist John Tyndall in the late 1850s, the Swedish physicist Svante August Arrhenius in the late 1890s, and the British engineer Guy Callender in the 1930s suggested a link between carbon dioxide concentration levels in the atmosphere and global warming, and speculated that human burning of fossil fuels was sufficient to change the climate. Empirical evidence showing increasing levels of anthropogenic produced carbon dioxide in the atmosphere began to be collected systematically in the late 1950s and 1960s by American scientists Roger Revelle, Hans Seuss, Gilbert Plass, and Charles David Keeling and others which, together with improved techniques for measuring the planet's climate, seemed to confirm the predictions of the early pioneers of climate science. Improvements in computing followed which provided scientists with better means to model future temperature changes if carbon dioxide concentrations continued to rise and what these would mean in terms of glaciation, sea levels, and desertification. Sceptics challenged this science, but over time, a scientific consensus has emerged that human activity has led to a build-up of carbon dioxide (and other greenhouse gases) in the atmosphere that is contributing significantly to climate change with potentially catastrophic consequences (Bolin, 2007; Weart, 2008; Hulme, 2009; IPCC, 2013).

Climate change policy has two prominent strands. The first seeks to bolster understanding of the causes and consequences of climate change by supporting the work of scientists. Governments provide funding and facilities to improve climate science. The second seeks to address the concerns raised by scientists through action to mitigate levels of carbon dioxide in the atmosphere or adapting to the consequences of a changing climate. Mitigation involves both measures to reduce emissions of carbon dioxide and efforts to remove the gas from the atmosphere. Adaptation involves a wide variety of measures designed to cope with the prospect of rising sea levels, desertification, violent storms, and other potential manifestations of climatic change. To count as climate change *policy*, research, mitigation and adaption efforts need to be *intentionally* adopted by government for that purpose. Government action to address a range of different problems may help promote research, reduce greenhouse gas emissions, or deal with the consequences of a warming world, but without a signal that such effects were intended they should not be regarded as examples of climate change policy. The US government financed research into weather patterns and atmospheric composition throughout the Cold War that boosted climate science, for example, but such action was undertaken for military and strategic reasons rather than to address climate change (Weart, 2008). Complicating matters is the fact that policy actors may not always signal their intentions clearly when making policy. Political expediency may mean they need to address climate change quietly rather than shout about their purpose. In such cases intentions become apparent after the policy action has taken place. Another complication is that policy actors may find ways to use old laws to combat climate change that were not originally intended for that purpose. Such action counts as climate change policy as it is the intention of the policy actor to use the law in a particular way rather than the intention of the original law-makers that matters.

Action to promote and evaluate research has dominated the response of the US government to concerns about climate change. Discrete federal funding for research into climate change, rather than general meteorological research or the by-products of military spending, began in the mid-1970s and has become a mainstay of the US government's approach to dealing with climate change despite periodic efforts by climate sceptics to cut such funding. Proposals to address the concerns raised by scientists have focused primarily upon mitigation strategies. Included among the policy tools suggested for addressing the problem are mandatory limits on emissions, carbon taxes, cap-and-trade systems of tradable permits, subsidies for renewable energy, and sponsoring research into carbon capture or sequestration. Another form of mitigation involves efforts to remove carbon dioxide already in the atmosphere. This typically involves action to plant trees and plants that can absorb atmospheric carbon dioxide though more radical forms of geoengineering have also been suggested. Mitigation has usually been viewed as morally superior to adaptation as the former purports to address the cause of the problem while the latter appears to condone the damaging behaviour that caused the problem

(Landy, 2010). Greater attention has begun to be paid to adaptation, however, as severe weather events such as Hurricane Katrina in 2005 and Hurricane Sandy in 2012 have illustrated the need to prepare against the possibility of rising sea levels and other climatic changes. Adaptation efforts include increased funding for flood defences, new infrastructure to deal with water shortages, and improved building standards (Farber, 2011).

Action to reduce levels of greenhouse gas emissions or adapt to the consequences of climate change can take place at all levels of government (Wolinsky-Nahmias, 2015). Many scholars have argued, in fact, that state and local governments have played a leading role in addressing the problem in the United States (Rabe, 2004; 2007; Byrne et al., 2007; Lutsey and Sperling, 2008; Selin and VanDeveer, 2009; Posner, 2010). The mitigation and adaptation efforts of these subnational governments are undoubtedly important to a comprehensive mapping of climate change policy in the United States. But given that the goal of this book is to trace and explain the development of US government policy, I have not attempted to describe the actions of lower level governments, except where these have driven policy at a national level. This focus can be justified not only by the need to correct the common idea that the US government has done little to address climate change, but also on the grounds of the importance of the US government in the political system. Although state governments have created regional compacts that include Canadian provinces and have occasionally signed agreements about climate change with foreign countries, the US government is *the* authoritative American actor in the international arena, has authority to impose policy solutions across the country, and can often pre-empt the authority of states to act. Explaining what drives US government action or, indeed, inaction is important given these circumstances.

The definition of climate change policy as intentional action to conduct research into climate change, mitigate greenhouse gas emissions or adapt to changes in the climate provides an essential starting point to tracking and explaining US government efforts to tackle the problem. By making clear what climate change policy involves, the definition helps identify what to look for when searching for policy, and equally important, what to ignore. Providing a definition of climate change policy, however, is only part of what is required to search for policy. Knowing *what* to look for is clearly vital to a study of US government climate change policy, but just as important is knowing *where* to look. The constitutional design and structural development of the US government complicate this search for climate change policy by providing multiple possible venues for action. Climate change policy can be made by the president through executive action and bureaucratic rule-making; Congress by passing laws and funding decisions; and the federal judiciary in court rulings. Within each branch of government numerous venues exist where climate change policy can be made. Units of the Executive Office of the President, parts of the federal bureaucracy, a considerable number of congressional committees, and different federal courts all have the potential to play a role in addressing climate change.

An important conceptual insight follows recognition of the policymaking capacity of different actors and institutions within the US government: multiple policy venues have the potential to produce climate change *policies* rather than a single clearly identifiable policy. These policies may be complementary or contradictory. Two prominent examples of the later are the Byrd-Hagel Resolution passed by the Senate in 1997 which undermined the Clinton Administration's negotiating position leading to the Kyoto Protocol, and the US Supreme Court decision in *Massachusetts* v. *EPA* (2007) which repudiated the Bush Administration's refusal to regulate carbon dioxide using the Clean Air Act of 1990. Presidential initiatives on climate change have also been affected by funding decisions taken by Congress on numerous occasions. Tracking and explaining the US government's efforts to address climate change, therefore, involves identifying a range of policy actions that may be moving at different speeds and sometimes in different directions. This complicates matters considerably as the policies need to be found and explained, but obviously leads to a fuller understanding of what the US government has done to tackle climate change.

US Climate Change Policies

US government action to address climate change began in the mid-1970s with congressional initiatives to promote research into the problem which led to the enactment of the National Climate Act of 1978, and President Carter's increasing concern about global environmental issues that culminated in publication of the *Global 2000 Report to the President* in July 1980. These early actions set in motion a process of research and reporting that kept the issue on the political agenda despite the indifference, if not hostility, of the Reagan Administration. The emergence of climate change as an international political issue in the mid-1980s also made it difficult for American politicians to ignore the issue completely. Major policy actions towards the end of the Reagan years included enactment of the Global Climate Protection Act of 1987 which strengthened research into the problem, and American support for the creation of the Intergovernmental Panel on Climate Change (IPCC). Both political parties in the 1988 elections made brief references to climate change in their platforms for the first time, and Vice President George H.W. Bush notably declared during the campaign that: "Those who think we are powerless to do anything about the 'greenhouse effect' are forgetting about the 'White House effect'" (Hecht and Tirpak, 1995, 383). Little change in policy, however, occurred during the Bush Administration. President Bush announced a US Global Change Research Program in 1990 to boost research into climate change, and Congress passed the Global Change Research Act of 1990 to give the Program a statutory basis. Measures to promote energy efficiency and develop alternative energy sources were included in the Energy Policy Act of 1992, but calls for more radical action to address climate change were vehemently resisted by the Administration despite pressure from Congress and foreign governments.

Although President Bush signed the United Nations Framework Convention on Climate Change at the Rio Earth Summit in June 1992, he insisted that mandatory greenhouse gas emission targets were not included in the treaty.

The election of President Bill Clinton in November 1992 initially promised a new era in climate politics in the United States. President Clinton had promised a new approach to climate change in the election campaign, Vice President Al Gore had been a leading advocate of action to address the problem since the early 1980s, committed environmentalists were appointed to prominent positions in the Administration, and the Democrats controlled both houses of Congress. Early enthusiasm to engage radically with the issue, however, soon dissipated. Clinton rejected proposals for a carbon tax, and the Administration's plans for a Btu tax failed to pass Congress. When the Clinton Administration published its climate action plan in October 1993 the emphasis was on voluntary measures to improve energy efficiency rather than new taxes or regulatory action to reduce greenhouse gas emissions. Republican victories in the midterm elections of 1994, which gave them majorities in both the House of Representatives and the Senate, effectively ended prospects of a major shift in domestic policy. Initiatives provided greater funding for research but little else with Congress opposing proposals to provide tax credits to promote energy efficiency. In the international arena President Clinton eventually decided to support proposals to include emission targets in a climate treaty despite overwhelming congressional opposition. In July 1997 the Senate passed on a 95–0 vote the Byrd-Hagel Resolution which stated that the United States should not commit itself to any limits or reductions in greenhouse gas emissions unless developing countries did the same. President Clinton signed the Kyoto Protocol in December 1997 that committed the United States to greenhouse gas reductions while exempting developing countries like China and India, but did not submit the treaty to the Senate for approval.

Opponents of action to address climate change viewed the election of President George W. Bush in November 2000 as an opportunity to reverse the previous Administration's policy initiatives. This faith appeared warranted when President Bush announced in March 2001 that the United States was withdrawing from the Kyoto Protocol. Bush claimed that the emission targets in the treaty lacked a scientific basis, would not achieve their objectives as developing countries could continue to increase their emissions, and would harm the American economy. Pressure for action, however, ensured that the Bush Administration could not reverse direction as much as it might have wished. In keeping with the approach of his two predecessors, President Bush launched initiatives to improve research, promote energy efficiency and the development of new technologies, and urged voluntary action to reduce emission levels. Democratic victories in the midterm elections of 2006, which gave them majorities in both houses of Congress, provided a boost to such efforts. The Energy Independence and Security Act of 2007 required automobile manufacturers to improve the energy efficiency of automobiles and other measures to promote energy efficiency, research into carbon sequestration, and funding to help developing countries tackle the problem.

A number of bills to reduce greenhouse gas emissions using a cap-and-trade system also began to be introduced and debated during this period though none were enacted. The biggest challenge to the incremental development of climate change policy during the Bush Administration came when the US Supreme Court ruled in *Massachusetts* v. *EPA* (2007) that the Environmental Protection Agency (EPA) had the authority to regulate greenhouse gases under the Clean Air Act of 1990. Although President Bush decided to ignore the ruling, the case provided the opportunity for a future radical change in policy without the need for a new law.

Barack Obama's victory in the 2008 elections promised a new direction in the US government's approach to climate change. Obama had stressed the need to address climate change during the election campaign, argued that a cap-and-trade system was the best way to reduce greenhouse gas emissions, and promised leadership on the world stage. Increased Democratic majorities in Congress served to bolster the sense that radical policy action was possible. Efforts to enact legislation to create a cap-and-trade system foundered in the Senate, however, as economic problems led Democratic senators to grow increasingly wary of supporting the measure. Republican victories in the midterm elections of 2010, which gave the Republicans a majority in the House of Representatives and reduced the Democratic majority in the Senate, effectively ended any prospect of *enacting* radical policy change. Faced with opposition in Congress President Obama employed administrative means to pursue his agenda. Using a mix of Executive Orders and Presidential Memoranda he set new fuel economy standards for automobiles and trucks, required the federal government to take measures to improve energy efficiency, and most significantly ordered the EPA to regulate greenhouse gas emissions from power stations under the Clean Air Act. The latter action was the first time that the US government had taken regulatory action to reduce emissions and represented a marked departure from existing policy.

A number of trends can be identified in this brief survey of US government climate change policies. First, spending on research and measures to improve energy efficiency and develop clean energy sources have dominated the government's policy responses. Calls for voluntary action, a faith in technological solutions, and a reliance on market solutions have formed the basis of these efforts to address the problem. Not until the Obama Administration did the government take significant regulatory action to reduce greenhouse gas emissions. Second, policies have developed incrementally with radical policy change limited to the Obama Administration. Climate change has not enjoyed the sort of "alarmed discovery" moment that produced major new laws like the Clean Air Act Amendments of 1970, the Clean Water Act of 1972, or the Comprehensive Environmental Response, Compensation, and Liability Act of 1980 (Downs, 1972). Finally, the ambition of the government's approach to climate change has been limited compared to many other environmental statutes. Whereas Charles O. Jones described the goals of the Clean Air Act Amendments of 1970 as "policy beyond capacity," American climate change policy over the last 40 years has clearly been well within capacity (Jones, 1975).

Explaining Climate Change Policies

Why has the US government responded to climate change in the way that it has? How can the shape and timing of various policy initiatives be explained? Answering these questions raises a litany of problems familiar to any student of public policy. Prominent among these difficulties are disputes about the appropriate level of analysis, arguments about agency versus structure, and concerns about how to isolate and measure variables. The multi-dimensional nature of climate change adds further complications to these traditional problems. With multiple venues capable of producing climate change policies the task is to explain how different political actors constrained by particular institutional settings respond to conflicting demands for action. This complexity means that it is impossible to develop a predictive model of policymaking that specifies exactly when and what climate change policies will be produced if certain conditions exist. Policymaking is more of an art than a science and is rarely amenable to the sort of precision required for scientific modelling. Faced with this reality, the best way forward is to provide a conceptual framework that allows some order to be imposed on the chaotic policy landscape that shapes the US government's responses to climate change. In this book I employ Kingdon's (2011) "streams approach" to achieve this task. Kingdon's framework provides a theoretically rich means to understand the various forces at play in explaining why the various venues of the US government have addressed climate change at particular times and why some solutions to the problem are considered while others are not (Pralle, 2009).

Previous studies have identified various factors to explain the US government's response to climate change (see Lee et al., 2001; Lutzenhiser, 2001; Christiansen, 2003). These range from broad cultural explanations that emphasise an ideological commitment to free markets to institutional explanations that focus on the structure of the political system to interest-based arguments that focus on the constellation of competing groups in the policy process. All identify important drivers of policy, but all have some shortcomings. Explanations based on general statements about American ideology or culture may identify a general disposition to act in a particular way but suffer from a lack of traction when it comes to explaining why something happened on a particular day. They also struggle to explain both contradictory policies produced by different parts of the political system and changes in policies over time. Institutional explanations suffer from a similar lack of traction in explaining the timing of policy action or changes in policy. They may explain how institutions privilege particular actors in the political system and mediate different policy views, but this emphasis on structure lacks the idea of agency that is necessary to explain the timing of policy production. Interest-based models have the explanatory traction lacking in many other types of explanation but usually fail to explain the origins of a particular constellation of interests acting within the political system.

Broad cultural explanations have been advanced in a number of studies to explain the US government's response to climate change (see Hoffman, 2015). These explanations typically stress the importance that an ideological predisposition to free market economics and limited government have in shaping attitudes and policy towards climate change, but also sometimes refer to a faith in technology as a solution to problems and an absence of the "precautionary" impulse found in some European countries as other important cultural factors driving policy. The role of economic thought in shaping climate change policies in the United States is summarised by Driesen (2010) who argues that the dominance of a neoliberal ideology over the last few decades has played a crucial role in determining how the problem is framed and the preferred choice of policy solutions. General acceptance of neoliberal ideas which stress the value of individual choice and free markets means a receptive audience for arguments claiming that government action to combat climate change undermines the individual freedoms at the heart of "the American Way of Life," and leads to a search for market-based methods to solve the problem. Another cultural explanation of the US government's approach to climate change emphasises the American faith in technology to overcome problems (Lee et al., 2001). The argument is that this faith has produced policies designed to promote new technologies at the expense of policies designed to change individual behaviour. Finally, Cass (2006) suggests that a reluctance to take "precautionary" action is an important cultural factor in explaining climate change policies in the United States. The argument is that action to address climate change has been shaped by the need for proven scientific evidence of harm rather than the possibility of harm. Connecting these various strands is a deep cultural belief in promoting economic growth and sustaining a particular way of life. Pielke (2011, 46) claims that this produces "an iron law of climate policy" where concerns about economic growth always trump efforts to reduce greenhouse gas emissions.

The various cultural factors that have been advanced to explain the US government's response to climate change point to a general predisposition to act in a certain way but lack the traction to explain the choice of particular policies by different institutions at specific times. To do this some account is needed of the role that various interests and institutions play in the formation of policy. One line of argument is that the power of vested interests has shaped the US government's climate policies. A common theme in much of the literature is the prominent place occupied by economic or business interests in the policymaking process (Rahm, 2009; Layzer, 2007; Gelbspan 2004, 1997). These groups have used their resources to question whether climate change is a problem and emphasise the costs of proposed mitigation efforts. Conservative think tanks have proved important allies in this effort (Jacques et al., 2008; McCright and Dunlap, 2003, 2000). Nordhaus and Shellenberger (2007) claim they have been helped in this task by the failure of environmental groups to develop appropriate strategies to sell the need for action on climate change (see also Bryner, 2008). Simply relying on interest group models to explain climate policies, however, neglects the role that institutions play in the policy process. Christoff and Eckersley (2011, 440), for example, argue that "the US political system displays many of the institutional features that are not conducive to robust climate policy." These include multiple veto points where opponents of action can block proposals.

Emerging from this brief survey of the main arguments that have been put forward to explain the US government's response to climate change is a sense that no single explanation captures what has happened. Cass (2006, 15) correctly points out that climate change policy "is a product of the state's institutional framework, domestic environmental policy norms, cultural differences, and historically contingent choices ..." The problem is that with so many different factors contributing to policymaking any effort to explain what is going on will rapidly become bogged down without a conceptual framework to impose some order. This is where Kingdon's (2011) "streams approach" helps. Kingdon argues that the US government will consider a problem in a particular way when the "problem," "policies," and "political" streams join to create a "policy window" for action. The "problem stream" is made up of the problems that policymakers believe worthy of attention. Given the large number of problems competing for the attention of policymakers the focus here is on problem definition, issue framing, and showing that one problem is more worthy of attention than another. The "policies stream" is composed of the potential solutions to a problem. Kingdon argues that ideas for dealing with a problem float around in a "policy primeval soup" evolving and mutating in a process similar to natural selection. Over time a "softening up" process may lead to some solutions gaining acceptance while others fail to convince policymakers of their worth. The "political stream" is composed of "such things as public mood, pressure group campaigns, election results, partisan or ideological distributions in Congress, and changes of administration" (Kingdon, 2011, 145). Changes in this "stream" can lead to changes in the way that problems and solutions are perceived. Opportunities for policymaking occur when something, perhaps a focusing event, leads to the "coupling" of these streams. Policy outputs will depend upon the content and way that the streams are coupled.

The role of "policy entrepreneurs" is an important part of Kingdon's model. Kingdon views "policy entrepreneurs" as "advocates who are willing to invest their resources—time, energy, reputation, money—to promote a position in return for anticipated future gain in the form of material, purposive, or solidary benefits" (2011, 179). Entrepreneurs may be elected politicians, bureaucrats, interest group leaders, or unofficial spokesperson for a particular cause (Cairney, 2012, 271). They play a central role in identifying problems, developing solutions, placing issues on the agenda of government, and persuading policymakers to act. In pursuit of their personal goals they help to couple the previously separated "streams." Kingdon argues that they "hook solutions to problems, proposals to political momentum, and political events to policy problems" (2011, 182). Policymaking may not occur without the presence of an entrepreneur. "Good ideas lie fallow for lack of an advocate. Problems are unsolved for lack of a solution. Political events are not capitalized for lack of inventive and developed proposals" (Kingdon, 2011, 182).

Kingdon's "streams approach" has a number of advantages as a conceptual framework for looking at climate change policy (Pralle, 2009; Hart and Victor, 1993). First, the strong emphasis given to problem definition, framing, and the evolution of solutions in the framework fits well with the intense debates that have raged over the last few decades about whether a problem exists, its causes, and what to do about it. Second, the "streams approach" can accommodate the fact that different approaches to dealing with climate change have emerged from different parts of the political system. Third, the ideas of evolution and change within the framework capture the path dependency of climate change policy (Knox-Hayes, 2012). Although primarily a framework for understanding agenda-setting and policy formation, Kingdon's "streams approach" also offers a way of explaining why particular policies are adopted. Kingdon argues that policy outputs depend on the content and mix of the "streams" when they are "coupled" (2011, 166). Important factors include the presence of policy entrepreneurs, the participants involved, public engagement with the problem, the solutions available, and the rules governing decision-making. An element of happenstance exists which adds a degree of unpredictability to policy outputs. Key political actors may be absent, for example, when vital decisions are made. The overall image conjured by Kingdon's framework is that policymaking is complex, chaotic, and messy. Parsimonious models of policymaking may strip away much of this messiness in the search for simple, predictive rules, but in so doing they remove the complexity that is fundamental to explanations of policy.

A discussion of historical context is fundamental to a conceptual framework based on Kingdon's "streams approach." The framework explains climate change policies in terms of how the problem is understood, the availability of solutions, and the configuration of the "politics stream" existing at a particular time. In this book I trace the development of climate change policy over the last four decades. Attention is paid to the role of policy entrepreneurs, changes in the way that the problem is defined, the rise and fall of solutions, and the consequences of changes in the staffing of government and the mobilisation of interests. Two government institutions figure prominently in this account: the Presidency and the US Congress. Both play a central role in shaping climate change policies. Presidents set agendas, suggest policy proposals, and their command of the executive branch provides a means to take action through orders and rule-making. Congress also identifies priorities, offers solutions to problems, and in the form of legislation makes authoritative statements of policy. The interplay, conflict, and occasional comity between these two powerful institutions are a major part of this book's narrative.

Plan of Book

This introduction is followed by a chapter which discusses and applies Kingdon's conceptual framework to climate change. The chapter identifies the salient features of each "stream" and explores the way they shape policymaking. Subsequent chapters provide the historical narrative to identify changes in the "streams" over time and the consequences for US climate change policy. The first of these chapters identifies the beginnings of political interest in climate change and maps policy developments to President Bush's signing of the United Nations Framework Convention on Climate Change in 1990. The remaining chapters are organised by presidential administration. Finally, the book concludes with a discussion of the dominant themes in US climate change policy.

Chapter 1

The Problem, Policies, and Politics

The factors that have shaped the US government's climate change policies over the last 30 years are numerous and complex. They include cultural and ideological forces, economic and political structures, the strength of competing interest groups, the partisan composition of government, and the actions of individual politicians (see Introduction). Without a conceptual framework to impose some order on this swirling mass of potential policy drivers the task of explaining the timing and nature of government action becomes next to impossible. The purpose of such a conceptual framework is to provide the means to understand why the US government has addressed climate change at particular times rather than others, and why it has chosen to do so in one way instead of another. Explaining agenda setting, policy formation, and policy choice are central to this task.

Kingdon's (2011) "streams approach" offers a way forward that meets most of these requirements (Pralle, 2009). Building upon earlier work by Cohen, March and Olsen (1972) Kingdon argues that the policy process is composed of three "largely independent" streams. The problem stream consists of the issues regarded as needing government attention. Kingdon notes that the list of potential problems that government could address is lengthy, and emphasises the importance of problem definition and issue framing in the competition for attention. The policies stream consists of the ideas or solutions that have been proposed to address problems. Kingdon argues that solutions evolve over time, and have a chance of survival if they meet several criteria. These include "their technical feasibility, their fit with dominant values and the current national mood, their budgetary workability, and the political support or opposition they might experience" (2011, 19–20). The political stream consists of factors such as the national mood, public opinion, election results, administrative and legislative turnover, and the prevailing constellation of interest groups. Kingdon argues that changes in these factors may lead to changes in the way that problems and solutions are perceived. Major policy change occurs when something, perhaps a focusing event or the actions of a "policy entrepreneur," brings the three streams together to produce a "policy window" when action is possible. Kingdon stresses that such "windows" do not remain open for very long.

In this chapter I follow Kingdon's lead and organise the climate change policy process into three streams. My purpose is to identify the salient features of each stream rather than map the changes that have occurred over time. Subsequent chapters will do that. The first section looks at climate change as a problem. I show that climate change has a number of characteristics that mean that it struggles in the competition for attention with other problems. The evidentiary basis of climate change allows room for disputes about whether the condition exists, permits questions about causal responsibility, and introduces uncertainty about consequences. Intense battles over framing rage as a result with different policy participants trying to define the problem to their advantage. The second section examines the solutions that have been proposed to deal with climate change and their fit with Kingdon's survivability criteria. I show that solutions such as technological fixes and tradable permit systems better fit these criteria than greenhouse gas emission limits, but that questions about cost and effectiveness dog most proposed responses to what is a global problem. The third section looks at the main elements that make up the political stream. I show that the configuration of political forces has not been amenable to major action to address climate change for most of the last thirty years or so. The period has been dominated by a neoliberal public philosophy, Republican control of key political institutions, economic problems, and wars that have diverted attention. I conclude the chapter by arguing that the conditions for a "policy window" to open have been few and far between, restricting opportunities for major policy change. The US government has adopted policies to address climate change, some of them important, but these have been primarily incremental rather than dramatic shifts in direction. The remainder of this book provides a detailed narrative to track these policy developments.

The Problem

The problem stream of the US government contains a large number of problems vying for the attention of decision-makers. Some of these problems have characteristics that demand attention. Unambiguous evidence of a condition or a situation that has adverse consequences for something most people care about is likely to attract attention. Other problems have characteristics that mean they struggle for recognition. The evidence of a condition or situation might be ambiguous and the existence of adverse consequences open to dispute. Climate change falls into this second category. Scientific evidence that anthropogenic climate change is occurring has taken time to assemble, and continues to be disputed by a few, raising questions over whether a condition exists. Further uncertainty surrounds predictions about the consequences of a changing global climate. Even when these predictions are relatively clear the consequences often lack immediacy. They seem to raise concerns for people living in other countries, or in the future, rather than contemporary Americans. These characteristics mean that issue framing is very important in the competition for the attention of decision-makers. Those who believe that government should take action have tried various ways of framing climate change to show that the problem is urgent, relates to other priorities such as economic growth and national security, and poses a threat to Americans in the short-term as well as future generations (Fletcher, 2009; Nisbet, 2009; Hale, 2010; Jamieson, 2011; Vezirgiannidou, 2013). Opponents of action contest these efforts with framing strategies of their own that emphasise uncertainty, costs, and the long-term nature of any consequences (McCright and Dunlap, 2000; Oreskes and Conway, 2010).

Identifying a Condition

Problems are conditions that people believe need to be addressed (Kingdon, 2011, 109). The starting point of the policy process, therefore, is the identification of a condition. Often conditions are readily apparent. Evidence of rush hour traffic congestion or rioting is difficult to dispute. Occasionally conditions may not be so obvious and need to be pointed out. Examples include poor education and child abuse. Climate change belongs firmly in this second category as the condition is impossible to experience directly and various indicators have to be used to show that it is occurring. Further difficulties are encountered when seeking to convince people that human activity is a cause of climate change as the relationship between cause and effect is also impossible to experience directly. Faith has to be placed in the findings of scientists rather than direct observation. Although scientific evidence that anthropogenic climate change is occurring has become increasingly robust over time, the nature of the evidential basis for identifying the condition continues to allow sceptics room to challenge the scientific consensus. Doubt can still be expressed about whether a condition exists.

Two questions need to be answered affirmatively to establish that anthropogenic climate change is a real condition. Is global warming occurring? And if so, is human activity the cause? Over the last fifty years or so a number of indicators have suggested that global temperatures are rising, that carbon dioxide and other greenhouse gases are accumulating in the atmosphere, and scientific models have been developed to link the two phenomena (Flemming, 1998; Weart, 2008). The answer to both questions posed above, in short, appears to be yes. Major scientific organisations in the United States and elsewhere have pronounced on a number of occasions that a consensus exists amongst scientists that global warming is occurring and that human activity is a significant cause (Oreskes, 2004; IPCC, 2013). The evidentiary basis of anthropogenic climate change appears to be clear. Not everyone is convinced, however, that the questions about global warming and causation have been answered quite so authoritatively. This is particularly the case in the United States where studies have suggested that there are more climate sceptics than in any other comparable country (Antilla, 2005; Demeritt, 2001). Disputes among scientists about the accuracy of data, the interpretation of trends, and the validity of climate models provide opportunities for these sceptics to question whether anthropocentric climate change is actually occurring (Booker, 2010; Carter, 2011; Montford, 2011; Fone, 2013; Vahrenholt, 2013). A significant problem is that climate science is not an experimental science that allows hypotheses about cause and effect to be tested under controlled circumstances (Rahm, 2009, 31). It is a science based on observation, measurement, and posited connections with room for disagreement about what is being measured, why, and what it all means.

The nature of climate change science affects public understanding of the issue. First, the absence of direct, immediate evidence of *global* climate change means that the public has to be persuaded that a condition exists that might need addressing. Individuals experience local weather conditions rather than a global climate, and this experience may be different from that posited by climate change scientists. Although measures of global average temperatures may show that the Earth has been warming over the last century, they hide the fluctuations in local weather that people actually experience (Hulme, 2009, 8–9). Talk of global warming sits uneasily alongside pronouncements of the coldest winter on record or shivering on beaches in summer. Second, claims that human activity is the cause of climate change similarly cannot be directly experienced. Individuals may witness smoke belching from factories or rivers the colour of rust, and make connections to incidents of illness or the ruin of a landscape even if they do not understand the pathologies involved. Making a connection from exhaust fumes to global warming is not so easy.

The primary indicators of global warming are measures of average air and sea temperatures covering the period from the mid-nineteenth century to the present day that have been constructed from individual thermometer readings taken around the world (see Figure 1.1). Although the various measures differ slightly in their methodologies, they all show a general warming over this period. They suggest that over the last 120 years the Earth's temperature has risen by about 0.8–1.0 °C with about two-thirds of this increase occurring over the last 30 years (NRC, 2011; IPCC, 2013). Techniques have also been developed to estimate global temperatures for eras lacking observational data. Some studies have "reconstructed" global mean temperatures back thousands of years using data derived from ice cores, trees, and marine sediments (see Singer and Avery, 2007, chapter 4). Further indicators appear to substantiate this picture of global warming. Evidence that glaciers, ice caps, and polar ice sheets have been retreating in most areas of the world has been used to confirm rises in global temperatures. Some studies have shown, for example, that by 2006 the Greenland and Antarctic ice sheets were losing 475 billion tonnes of ice per year (Black, 2011).

Scientists have advanced two broad explanations of climate change over the last century. The first type of explanation claims that periods of global warming (or cooling) are natural phenomena. Changes in the Earth's orbit, solar activity, meteorite impacts, volcanic activity, and the movement of tectonic plates have all been posited as causes of climate change at one time or another (Flemming, 1998). Some of these events are cyclical while others are essentially random. Glaciers may advance and retreat on a regular basis over millennia because of known variations in the Earth's orbit, but climatic changes caused by volcanic eruptions or meteorite impacts are abrupt and largely unpredictable. The second type of explanation places much of the blame for recent global warming on human activity. Massive increases in mankind's burning of fossil fuels over the last 150 years has led many scientists to argue that sufficient carbon dioxide has been released into the atmosphere to create a "greenhouse effect" (Flemming, 1998; Weart, 2008; Hulme, 2009). Carbon dioxide and other "greenhouse gases" reduce the amount of heat escaping from the Earth thereby raising temperatures. Other human activities such as deforestation and clearing land for farming may also contribute to a lesser degree to this warming (Pielke, 2010, chapter 1).

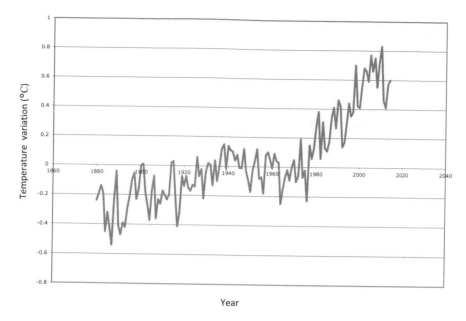

Figure 1.1   Global surface air temperature anomaly
Source: NASA/GISS.

These two broad explanations of climate change are not mutually exclusive, as global warming could be caused by both natural and anthropogenic factors. Changes in the Earth's orbit *and* increased concentrations of carbon dioxide in the atmosphere, for example, might explain temperature rises. But they cannot both be the *primary* cause of a changing climate. Successive reports by the United Nation's Intergovernmental Panel on Climate Change (IPCC) have identified human activity in increasingly confident terms as the main cause of recent global warming (Bolin, 2007). The cautious claim made in the IPCC's *Second Assessment Report* published in 1996 that "the balance of evidence suggests that there is a discernible human influence on global climate" had changed to the far more assertive statement that "It is *extremely likely* that human influence has been the dominant cause of observed warming since the mid-20th century" in the summary for policymakers of the IPCC's *Fifth Assessment Report* published in 2013 (Hulme, 2009, 51; IPCC, 2013). Not all scientists agree with the IPCC's conclusions. Some continue to claim that natural causes are far more important in explaining any rise in global temperatures (Plimer, 2009; Carter, 2010).

Disputes among scientists about methodologies and interpretation of results are a fundamental part of the scientific process (Hulme, 2009, 75). Climate scientists have argued about matters such as the possibility that "urban heat islands" have distorted the evidence of global warming, whether the spread of weather stations is adequate to measure global temperatures, and the effectiveness of tree rings as indicators of historical temperatures. These disagreements have been seized upon by sceptics of climate change to deny that there is a scientific consensus about global warming (Horner, 2008; Booker, 2010; Carter, 2011; Montford, 2011; Fone, 2013). They suggest that organisations such as the IPCC have deliberately overstated the consensus among scientists about climate change and sought to censor dissent (Montford, 2012; Ball, 2014). The interpretation of data has also been questioned. Climate sceptics note that graphs of temperature changes over the last century or so actually show a period of cooling (the Little Cooling) from the 1940s to the 1970s, and that measures of temperature over the past millennium reveal that the Earth is currently cooler than at the peak of the Medieval Warm Period between 1100 and 1200. Disagreements among scientists about the workings of the carbon cycle and measurements of climate sensitivity, together with data which shows falls in global temperatures in some years despite continued increases in carbon dioxide concentrations in the atmosphere, have also been used by sceptics to question the idea of anthropogenic climate change. Natural phenomena rather than human activity, they assert, are responsible for any global warming that might (or might not) be occurring (Plimer, 2009; Carter, 2010).

Scope for challenging the existence of anthropogenic climate change has remained despite the advances in climate science that have taken place over the past 50 years. This ability to cast doubt on whether a condition exists has implications for climate policy as it may undermine public support for government action. Even the provision of unchallengeable evidence that climate change is occurring, however, would not automatically lead to the condition being viewed as a problem. Defining climate change as a problem requires both agreement that the condition exists *and* acceptance that the condition has adverse consequences that need to be addressed.

The Consequences of Climate Change

Identifying a condition is a necessary but not sufficient part of establishing that a problem exists. People have to deal with conditions all the time, and may even refer to them as "problems" in everyday conversation, but the condition itself is not a problem in a public policy sense (Kingdon, 2011, 109; Stone, 1988; Rochefort and Cobb, 1994). To be defined as a problem, a condition needs to be interpreted as something that has adverse consequences that warrant government action to change the condition or alleviate its effects. Climate change fares poorly in this process. Although the mooted consequences of climate change often suggest catastrophe, details of what will happen, when, and to whom have frequently been imprecise and lacking in immediacy. This affects climate policy in two important ways. First, uncertainty about the consequences of climate change provides an opportunity to question whether a problem exists. This can involve challenging the projections of future climate change or disputing the interpretation of what those changes mean. Second, a perceived lack of immediacy may lead to acceptance that there is a problem, but low ranking in terms of priorities. Opponents of government action may argue that other problems that clearly effect people today are more deserving of attention than climate change (Lomberg, 2007).

Controversy has surrounded efforts to predict the consequences of climate change. Authoritative answers to questions such as what will happen if global temperatures continue to rise and when will the effects be felt have been difficult to provide, because they depend upon making predictions about the future based on an incomplete understanding of the way that the global climate works, estimates about levels of future economic activity, uncertainty about how any climatic changes will affect ecological, agricultural, economic, and social systems, and judgements about the threat posed by such changes. The global climate is a vast, complex and chaotic system that defies easy modelling (Hulme, 2009, 83). Factors such as changes in the composition of the atmosphere and the behaviour of oceans, interactions between these two systems, the potential existence and effects of "feedback loops," and the possibility of "tipping points" that lead to abrupt changes in conditions all need to be taken into account when constructing climate models. Although scientific understanding of many of these factors has improved over time, there is still much that is not known. This has led to disagreement about such things as the pace of future global warming, the speed that polar ice, glaciers, and snow packs will retreat or disappear, the extent of potential desertification, and the size of any rise in the sea level. Further uncertainty surrounds the consequences of any future climatic changes on ecological and human systems. Will global warming lead to the loss of valuable agricultural land or increase the acreage that can be cultivated and lengthen the growing season? Will tropical diseases spread to current temperate zones? Will coastal cities disappear under sea water? And will the consequences be felt by Americans or people living in far away places that most Americans have never heard of? Climate science has few undisputed answers to these questions.

Predictions of the rate of global warming depend on estimates of future greenhouse gas emissions and how "climate sensitivity" to these emissions is modelled. An indication of the difference such factors can make can be seen in the projections of global temperature rises found in the IPCC's *Fifth Assessment Report* (2013). The IPCC predicted that global temperatures might be between 1.5 °C and 4.0 °C higher in 2100 than in 1900 depending upon the scenario modelled and particular assumptions made about climate–carbon cycle feedbacks. Temperature rises in the lower range assume that greenhouse gas emissions will stabilise around 2000 levels while those in the upper range assume growth in emissions and stronger climate–carbon cycle feedbacks. Most climate scientists agree that warming of the oceans, for example, will reduce absorption of atmospheric carbon dioxide but do not agree on the strength of this feedback loop. Other studies have suggested that temperatures might rise more than these estimates. One study even claims that global temperatures might rise by as much as 11.5 °C over the century if emissions of carbon dioxide double (Stainforth et al., 2005). Temperature rises of this size, however, are regarded as extremely unlikely (Monbiot, 2007).

Uncertainty about future temperature rises complicates discussions about the consequences of global warming as a range of scenarios have to be modelled. The IPCC's *Fifth Assessment Report* (2013) contains an estimate that increases in global temperatures over the next century will probably lead to sea level rises of between 0.26 m and 0.98 m, but admits that scientific understanding of the factors driving rises in the sea level are not fully understood and there is a possibility that sea level rises might be higher. Similar variation exists in a number of the IPCC's other predictions about the consequences of global warming. Estimates of the reductions in Arctic sea ice by 2100 range from 43 per cent to 94 per cent, while predictions about the loss of global glacier volume outside Antarctica range from 15 per cent to 85 per cent. Loss of near-surface permafrost at high northern latitudes is estimated to be between 37 per cent and 81 per cent. According to the IPCC, global warming during the twenty-first century is "virtually certain" to cause "more frequent hot and fewer cold temperature extremes over most land areas," "very likely" to produce more and longer heat waves, and "likely" to increase the area encompassed by monsoon systems. Scientific understanding of climate change has undoubtedly improved over time but the variation evident in these predictions is testimony that much remains unknown.

What do these general trends mean for people living in the United States? Are they something to be concerned about or parked for consideration on another day? A 2009 report by the US Global Change Research Program stated that the impact of climate change could already be felt in the United States (USGCRP, 2009). Claims that average temperatures in the United States have risen by 1.1 °C over the last 50 years; precipitation has increased by 5 per cent; extreme weather events have become more frequent; glaciers have retreated rapidly; and sea levels have risen along the Atlantic coast are among those made in the report. The report's efforts to predict what would happen in the future, however, reveals similar uncertainty to that found in the IPCC's predictions about global changes. The report contains a number of predictions about the likely impact of climate change on the United States. Like the IPCC's estimates of future climatic changes, they provide different scenarios based on levels of greenhouse gas emissions. The report estimates that average temperatures in the United States will increase by 1.1–1.6 °C by 2029 and by 2.2–6.1 °C by the end of the century, with regional variation further complicating the picture. Northern areas are predicted to become wetter and southern areas drier with increased periods of drought likely in the southwest. Melting icepacks and glaciers will lead to likely increases in sea levels of 0.6–0.9 m in New York City and 1.04 m in Galveston, Texas, by 2100.

Persuading people that these projections about the impact of climate change are a cause for concern has proved problematic. Not only are they clouded in uncertainty, appearing to be something that will happen in the future, they also lack meaning. What does a further temperature rise mean to the lives of people in the United States? Will the advantages of longer growing seasons and shorter transport links because of melting Arctic sea ice outweigh the costs of water shortages in some parts of the country and coastal flooding in others? Should Americans worry if a remote Pacific island disappears under the waves in 50 years' time? The difficulty of answering such questions has created a framing battleground in which policy actors attempt to convince decision-makers and the public of the need (or otherwise) of government action. Proponents of government action have sought to frame the problem in ways which stress that the consequences of climate change will harm Americans and require urgent attention. Such frames have included efforts to define climate change in terms of public health, energy security, and human rights (Fletcher, 2009; Nisbet, 2009; Hale, 2010; Jamieson, 2011; Vezirgiannidou, 2013). The Pentagon has even framed climate change as a national security risk (USDoD, 2014). Opponents of government action have attempted to frame the issue in terms of scientific uncertainty, economic costs, and the long-term nature of any threat (McCright and Dunlap, 2000; Oreskes and Conway, 2010; Oreskes, 2011).

The struggle to frame climate change in terms that provide meaning to scientific predictions about the consequences of a warming planet indicates profound difficulties in defining the problem. This means that climate change typically loses out in the competition for the attention of decision-makers. The condition may be recognised but the consequences regarded as uncertain and happening in the future. Other problems often seem more obvious and immediate. Even when attention has been given to climate change, perhaps because of the publication of a UN report or a focusing event like a hurricane, disputes about what can and should be done create further opportunities to engage in a contest over framing.

Policies

Numerous potential responses to climate change exist in the policy stream (see Helm, 2012; Lomberg, 2007). Kingdon argues that a process similar to natural selection determines which of these responses will be taken seriously by decision-makers and which will be ignored (2011, 116). The "criteria for survival" identified by Kingdon include judgements about whether the response will work, its fit with society's values, its costs, and the likely public reaction to its adoption. Proposals to address climate change often appear to fare poorly when measured against these criteria providing opportunities for opponents to argue against their adoption. The global nature of climate change and disputes about science allow questions to be asked about the effectiveness of action to reduce greenhouse gas emissions. Many policies appear to have a notion of "sacrifice" at their centre which is at odds with dominant American values that stress materialism and economic growth (Maniates and Meyer, 2010). The costs of taking action seem high particularly when assessed against uncertain future benefits. And public support for action is often not clear cut. Proponents of government action to address climate change have sought to counter these arguments by framing policies in ways to make them more acceptable (Fletcher, 2009). Examples include arguments that action to address climate change will promote energy security and boost employment in the "green economy" (Vezirgiannidou, 2013). The result has been a fierce framing contest over climate policies that mirrors that found in arguments over whether the problem exists.

Potential Solutions

Problems have little chance of being addressed seriously by government without some idea of what a solution would look like (Kingdon, 2011, 178–9). Knowing what could be done to change the underlying condition or ease the consequences resulting from that condition is essential if a policy response is going to be meaningful rather than symbolic. The difficulty with climate change is that the problem appears to defy easy solutions. First, doubts have been raised as to whether a solution to anthropogenic climate change is possible. The IPCC's *Fifth Assessment Report* (2013, 25) notes that: "Most aspects of climate change will persist for many centuries even if emissions of CO_2 are stopped." Lower levels of warming are the best that might be achieved if emissions are reduced the report concludes. Second, the effectiveness of action by a single country, even one as important as the United States, may be limited if other countries increase their emissions. Third, identifying what to do to ease the consequences of climate change is difficult given the uncertainty surrounding those consequences. These doubts mean that the starting point in the search for solutions is often scepticism about what, if anything, can be achieved by government action. When concerns about costs, economic growth, and behavioural changes are added to the mix the prospect of finding an acceptable solution becomes even more remote. Climate change is a "wicked," or even "super wicked," problem that defies easy resolution (Rittel and Webber, 1973; Lazarus, 2009; Levin et al., 2012).

Promoting research into the causes, effects, and means of addressing climate change has been frequently advanced in the United States as a response to the problem. Although difficult to define as a solution to climate change and sometimes criticised as a means of delaying concrete action, such research can have an impact on the policy process. Not only will the studies mandated by proposals for more research serve as further indicators that a condition exists, they may also help institutionalise attention to the problem within government (Downs, 1972; Peters and Hogwood, 1985). A requirement to present research findings to federal bureaucrats or congressional committees might help limit any decline in attention to the issue. Recognition of these two points means that climate sceptics have often opposed proposals to promote research. Research might appear a relatively benign and cheap policy option, but problems occur when the results of that research show things that some policy actors would rather remain unknown. The Bush Administration (2001–2009) tried to censor the findings of government climate scientists, for example, when those studies challenged the official line on climate change (Mooney, 2006; Bowen, 2009; Bradley 2011).

Proposed solutions to climate change can be divided into two different approaches: solutions designed to *mitigate*, or reduce, levels of carbon dioxide in the atmosphere; and solutions designed to *adapt* to the consequences of climate change. Mitigation includes both proposals to reduce emissions of greenhouse gases (or stop other human activity that might cause climate change) and measures to remove the gases from the atmosphere. The range of mitigation strategies designed to reduce emissions includes traditional command-and-control methods of standard setting, cap-and-trade systems of tradeable permits, carbon taxes, subsidies for renewable energy, and promoting the development of new technologies to capture or sequester carbon. Proposals to remove greenhouse gases from the atmosphere range from planting more trees to using chemicals to absorb atmospheric carbon dioxide. Pielke has called these methods "carbon remediation" (2011, 133). Adaptation involves proposals designed to ameliorate the consequences of climate change. These range from radical geoengineering projects that seek to cool the planet by manipulating the atmosphere to reflect more sunlight back into space to local improvements to sea defences. The distinction between mitigation and adaptation is "blurry" at times with some solutions having the potential to reduce greenhouse gas emissions and help cope with climatic changes (Farber, 2011, 480). Improved insulation, for example, would reduce energy usage and improve protection against the consequences of climate change.

A number of proposals to mitigate levels of carbon dioxide have been suggested over the last four decades (see Helm, 2012; Rahm, 2010; Helm and Cameron, 2009). These can be placed into broad groups according to their scope and regulatory approach (Jordan et al., 2011). The scope of mitigation plans has ranged from the global to the domestic. Some have advocated negotiating international agreements to reduce global emissions of greenhouse gases or geoengineering projects to remove carbon dioxide from the atmosphere. Others have proposed setting economy-wide emission standards. Another approach has been to suggest reducing emissions from prominent sectors of the economy such as power generation and transport. And others have focused on reducing the carbon footprint of families and individuals. The regulatory approach to achieving these goals has also varied considerably. Some proposals mandate emission levels, or the use of particular technologies, and establish criminal penalties for those who fail to meet these requirements. Others suggest employing market mechanisms to create incentives or disincentives to reduce emissions (Harrison et al., 2011). These include establishing markets and trading carbon dioxide permits, imposing taxes on carbon dioxide emissions, and providing subsidies to encourage the development of new technology or promote changes in behaviour. Finally, some argue that encouraging voluntary action to reduce emission levels is the best way forward. Each proposal has advantages and disadvantages when judged against Kingdon's "criteria for survival" depending on the mix of scope and approach being advocated, but none fare particularly well.

Negotiating an international agreement that places limits on global emissions of greenhouse gases is commonly regarded as the best solution to climate change. Climate change is a global condition with global consequences that requires global action to reduce atmospheric concentrations of carbon dioxide and other greenhouse gases. A number of difficulties, however, plague international action as a solution to climate change. Some of these difficulties are generic; others are particular to the United States. Overall they cast doubts on the "technical feasibility" and fit with the country's dominant values. The difficulty of negotiating an international agreement starts with establishing an emissions target. Scientific uncertainties and normative judgements dog this task. Not only does it require determining what temperature rises are associated with particular emission levels, but also involves a judgement about what level of warming is acceptable and what is dangerous (Steffen, 2011, 29; Pielke, 2011, 148). Next is the task of actually reaching agreement on how the emission target will be achieved. This involves securing the agreement of countries with vastly different levels of economic development and energy needs about who will do what and when. Fears that the United States is being asked to do too much while other countries are getting away with doing very little often arise at this stage. Concerns about fit with American values supplement these doubts about "technical feasibility" to undermine support for international solutions to climate change in the United States. The United States has historically been wary of involvement in international organisations unless ceded a degree of control over institutions and the decision-making process (Lee et al., 2001, 385). The idea of giving an international organisation authority to compel compliance with emission targets is anathema to most Americans and easily characterised as an attack on the country's sovereignty.

Proposals to remove carbon from the atmosphere using simple geoengineering techniques such as planting more trees fare well against the "criteria for survival." Expanding and protecting forests raises land-use questions, but the "technical feasibility" is clear, costs are relatively low, and limited behavioural change is required from Americans. More radical geoengineering proposals that involve using chemicals to remove carbon dioxide from the atmosphere or reflect more sunlight back into space fare less well against the "criteria for survival." On the one hand, they play well to the American faith in technological innovation to solve problems and require no change in behaviour, but the "technical feasibility" and governance questions raised are considerable. Some scientists question whether such techniques will work and worry about the risk of things going wrong (Hulme, 2014). Further questions arise over who should take the decision to use such techniques, and who will take the blame if something goes wrong (Burns and Strauss, 2013; Victor, 2009; Rahm, 2009, 160). If some form of collective decision-making is needed to answer these questions many of the difficulties associated with negotiating international treaties reappear.

Setting domestic emission standards avoids the transaction costs associated with negotiating with other countries and has been proposed as a means of addressing climate change on a number of occasions. Proponents argue that a reduction in American emissions would have a tangible effect on atmospheric levels of greenhouse gases as the United States is one of the largest producers of carbon dioxide in the world. Such a solution, however, has been questioned in terms of "technical feasibility," costs, and fit with the country's dominant values. Some of these concerns mirror the generic difficulties of determining emission levels encountered in proposals for international action. What emissions target will produce the required slowdown in global warming? Further questions about effectiveness arise if American reductions in emissions are offset by rises elsewhere. Critics argue that American action will count for little or nothing if countries like China and India increase their emissions. Questions about cost follow with critics claiming that unilateral action involves an economic cost not borne by other countries and places the United States at a disadvantage when compared to its competitors. Finally, concerns about the level and means of government intervention in the economy that will be needed to meet emission targets have been expressed. Historically, Americans have resisted government intervention unless confronted with a clear crisis (Lee et al., 2001, 384). Doubts about the existence and consequences of climate change undermine perceptions that the country is facing a crisis that requires government intervention on the scale needed to reduce greenhouse gas emissions. Cultural and ideological attitudes towards government also mean that certain regulatory approaches have a better chance of adoption than others. Market-based approaches to addressing the problem such as cap-and-trade have been viewed more favourably than traditional command-and-control methods of regulation (Rahm, 2010, 136–8; Driesen, 2010). Proposals for carbon taxes have also struggled for acceptability in an anti-government, anti-tax, environment (Rabe, 2010).

Sector-based solutions that focus on major producers of greenhouse gases such as power generation and transportation have a number of advantages when measured against the "criteria for survival." Power generation and transportation produce just over 60 per cent of greenhouse gas emissions in the United States (see Figure 1.2). suggesting that significant reductions in these two sectors would have a meaningful impact on overall emission levels although the impact on global emission levels is open to debate (USEPA, 2014). Proposals to reduce emissions from power generation and transport have advanced a number of technological fixes. These include promoting the development and use of renewable energy sources, creating a "smart grid" that will distribute electricity more efficiently, employing methods of carbon capture and sequestration, establishing improved fuel economy standards for automobiles, and investing in new forms of public transportation. The stress on technology in these proposals accords well with American cultural values, mandates no changes in individual behaviour, and has been a dominant strategy in most environmental laws enacted since the 1970s. Government support for technological development in the form of grants, subsidies, and tax breaks may also bring economic benefits. Sector-based solutions that focus on power generation and transportation fare less well against cost criteria. Doubts about the government's ability to finance a programme of technological development may be raised in periods of budgetary stress, and individual Americans may baulk at

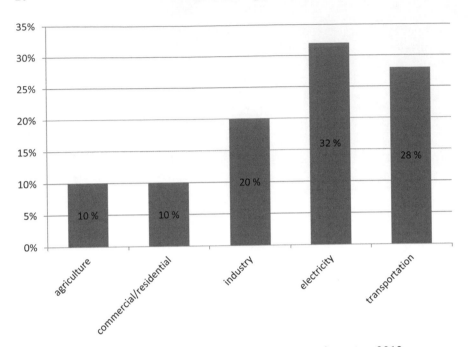

Figure 1.2   US greenhouse gas emissions by economic sector, 2012
Source: USEPA (2014) "Inventory of US Greenhouse Gas Emissions and Sinks 1996–2012,"
Washington, DC, 15 April.

paying more for energy and motoring. Support for public transportation has also
been limited in a country with a deeply ingrained "car culture" and cities designed
around the use of an automobile (Rajan, 1996).

A final raft of solutions focuses on reducing the carbon footprint of
communities, families, and individuals. These range from proposals to mandate
improved energy efficiency in buildings or consumer products to efforts to persuade
individuals to change their behaviour. Proponents argue that small changes
in individual energy consumption can result in significant reductions in overall
energy use. Individual consumption of energy has been estimated to account
for about 40 per cent of the country's carbon dioxide emissions (Szasz, 2011).
Doubts have been expressed, however, about the acceptability of government
mandates to improve energy efficiency, and the effectiveness of campaigns to
persuade consumers to change their behaviour. Proposals for mandates often run
into industry opposition, and run counter to the anti-regulation ethos that has
prevailed in the United States since the late 1970s, while campaigns to create
a "green consumer" depend on individuals believing that climate change is a
problem, and the cost of changing behaviour. Individuals will not change their
behaviour voluntarily if they do not believe a problem exists or if the costs of
changing their consumption is deemed too high.

Solutions designed to adapt to climate change have often played "second fiddle" to mitigation strategies. Proposals to reduce greenhouse gas emissions have been regarded as addressing the cause of the condition whereas proposals to adapt to the consequences of climate change have been seen as condoning the activities that cause the condition (Landy, 2010). However, the realisation that the obstacles to agreeing significant reductions in greenhouse gas emissions may be too great to overcome, and that any such cuts may be insufficient to reverse climatic changes already underway, has led to growing interest in adaptation strategies. A wide range of ideas have been suggested about the action that needs to be taken to deal with a changing climate. These include genetically engineering crop varieties that are more tolerant of heat and drought, building better coastal defences, developing new land management techniques to protect ecosystems, and using water more efficiently. Many of these ideas have been welcomed as ways of improving living conditions in general, but questions have been asked about their effectiveness and cost. A significant concern derives from the uncertainty surrounding predictions about the consequences of climate change at the local level. Will adaptive strategies deal with the change in conditions that will actually arise in the future? Questions about feasibility may arise even with increased confidence about the consequences of climate change. Can improved coastal defences really protect communities if sea levels rise considerably?

Selling Solutions

The ability of climate change sceptics to highlight areas of scientific uncertainty, economic costs, and lack of fit with dominant cultural norms presents a challenge to those who believe that government needs to take action to address the problem of a warming planet. Selling solutions to the public and political elites is difficult even when climate change is recognised as a problem. Two potential strategies exist to meet this challenge. First, proponents of action can "engage" with the arguments presented by opponents (Jerit, 2008). They can contest claims of scientific uncertainty, high costs, and lack of effectiveness, or seek to show that certain regulatory approaches are consistent with neoliberal thinking and other cultural features. The difficulty with this strategy is that debate takes place on ground chosen by opponents of action. Second, those who believe that government should address climate change can seek to frame potential solutions in ways that broaden their appeal. A common way of doing this is to argue that dealing with climate change will also solve other problems that Americans might regard as more important or urgent (Vezirgiannidou, 2013). Claiming that greater use of renewable energy will lessen the country's dependence on imported oil is one example of such a framing strategy. The purpose is to stress the co-benefits of action to tackle climate change and move away from arguments that emphasis economic costs or sacrifice (Peterson and Rose, 2006). Both strategies try to improve the performance of proposals to deal with climate change when measured against Kingdon's "criteria for survival."

Countering claims that action to address climate change is "technically unfeasible," involves high costs, will lead to greater government bureaucracy, or produce other unwanted consequences, requires proponents to engage in debate on grounds chosen by their opponents. This means that they will often have to discuss topics that they might wish to avoid given the chance. Such engagement can help establish public legitimacy for proposals if the arguments of opponents are addressed and rebutted, but the challenge is considerable as making claims about problems or costs is usually easier than proving they are exaggerated or non-existent (Jerit, 2008; Riker, 1996). Concern that discussing problems and costs merely reinforces public perceptions of the sacrifices associated with dealing with climate change has increasingly led proponents of action to frame solutions in terms of the wider benefits they will bring to the country. One way of doing this is to frame action on climate change as boosting economic growth. The "green economy" frame claims that action to address climate change will promote economic growth through innovation, infrastructure development, and the creation of "green jobs" (Fletcher, 2009; Bowen and Frankhauser, 2011). Another way of reframing solutions is to argue that action on climate change will provide a solution to America's longstanding problem of energy security (Bang, 2010; Greene, 2010). Proponents claim that promoting "green" sources of energy will reduce the country's reliance on imported oil.

The effectiveness of reframing strategies as a means of selling solutions to climate change has been questioned on a number of grounds (Vezirgiannidou, 2013). First, a concern with costs reappears when discussion moves beyond general statements about the benefits of a "green economy" or energy security to look at concrete proposals. Greater use of renewable energy usually leads to consumers paying more and increased burdens for suppliers. Second, alternative ways of achieving economic growth and energy independence are available. Stressing energy supplies and economic growth may invite proposals for increased drilling in national parks, greater use of nuclear power, and measures that boost the economy in ways that increase the production of greenhouse gases. Finally, the reduced emphasis on climate change in such frames may lead to a feeling that the issue lacks importance. The dominant message is that economic growth or energy security is more important than climate change. Paradoxically, this may make it more difficult to sell solutions to climate change. No proposal to address climate change is cost free and without sacrifices of some form. The need is to persuade the public and decision-makers that such costs and sacrifices are worthwhile. This may become more difficult if frames are used which suggest that climate change lags behind other concerns in importance.

The challenge for proponents of government action to deal with climate change is not a lack of options but selling those options to the public and decision-makers. Potential ways of addressing climate change are plentiful. The policy stream contains numerous ideas about how to mitigate or adapt to climate change. These range from solutions that require global action to those that focus on individual behaviour, and include the full gamut of regulatory approaches. Uncertainty about the causes and consequences of climate change, and disputes about the "technical feasibility" and costs of action, however, make selling solutions difficult. Some have characteristics that make them easier to sell than others, but all suffer from an uncertain demand for what is being offered. Although events and changes in the "political stream" may boost demand at particular times, both the public and decision-makers have often failed to identify climate change as a priority and that has made selling solutions difficult.

Politics

The way that problems and solutions are viewed is dependent upon a range of factors, including "such things as public mood, pressure group campaigns, election results, partisan or ideological distributions in Congress, and changes of administration," that constitute the "political stream" (Kingdon, 2011, 145). These factors determine who makes policy decisions, the political context in which they operate, and the considerations they will assess when evaluating whether to address a problem and what solution to choose. Over the last forty years or so the nature of the "politics stream" has generally proved hostile to action on climate change. The public mood has been dominated by a neoliberal sentiment characterised by a general hostility to government regulation; public opinion has often been ambivalent about whether climate change is a problem that needs to be addressed as a priority and uncertain whether the costs of action are worthwhile; Republicans have occupied positions of institutional authority for most of the period, giving them power to shape the agenda; and energy groups have used their wealth and influence to campaign aggressively against action to address climate change. Only for short periods have changes in the "politics stream" produced conditions receptive to viewing climate change as a significant problem and to giving radical solutions serious consideration.

The Public Mood

Kingdon uses the terms "public mood," "national mood," "climate in the country" interchangeably to describe the "common lines" of thinking about such things as the state of the country, well-being of society, and attitudes towards government that pertain at a particular time (2011, 146). Not to be confused with public opinion, the "public mood" is something that policymakers "sense" from a variety of sources, including election results, media reports, town hall meetings, and the emergence of social movements. Whether this perception of a "mood" corresponds with what the mass public actually think is beside the point—what matters is that policymakers believe that it does. Kingdon argues that policymakers' "sense" of the "public mood" will shape both their willingness to address purported problems and the solutions that will be favoured. When the "public mood" favours an enhanced role for government in dealing with economic and social problems, for example, proposals requiring government spending and regulation will be viewed more positively than when the reverse is true. The "public mood," in short, "makes some proposals viable ... and renders other proposals simply dead in the water" (Kingdon, 2011, 149).

The "public mood" in the United States over the last thirty years or so has not been favourable to action on climate change. Most commentators have argued that the country has been dominated by a neoliberal sentiment that stresses free markets, limited government, de-regulation, and low taxes for most of this period (Mirowski, 2014; Turner, 2008; Harvey, 2005). The dominance of this backdrop has shaped the way that proposals like those dealing with climate change that require increased government regulation have been received, and also offered opponents of action ways of framing debate to their advantage. Claims that proposals to address climate change will lead to "Big Government" have had a strong resonance, and have forced proponents of action onto the defensive. Periods when a change in the "public mood" have created a more amenable environment for consideration of action on climate change have been few and far between. The election of President Barack Obama in 2008 appeared to presage a change in the "public mood" more favourable to government action to address a range of social and economic problems, but this "mood" did not last long before the underlying neoliberalism reappeared.

Public Opinion

Public opinion on a particular subject *helps* policymakers determine whether a problem is worthy of consideration and what action might be acceptable (Erikson and Tedin, 2014). This is not to say that policy simply follows the lead of opinion polls, but to point out that what people think about problems and solutions will be taken into account when policymakers consider what to do and when. Public concern about a problem is more likely to generate a response from policymakers, for example, than public indifference. Recognition of this fact often produces intense campaigns to shape public opinion to further particular causes. Proponents of action to do something about a problem will seek to persuade the public that a condition exists which has adverse consequences that need to be addressed as a matter of urgency while opponents will seek to do the opposite. Winning this battle to shape public opinion may not necessarily guarantee that policymakers will respond in the desired way as other factors may influence their action, but showing that the public has a clear view of a problem is an important input into the policy process.

The fact that climate change cannot be directly experienced means that public understanding of the issue is shaped by media coverage of scientific findings and natural events. Newspapers, television and radio news, and other forms of popular communication provide the primary means through which the public learns about climate change (Wilson, 1995). The media essentially translates the professional language of science into narratives and images that people can understand (Boykoff, 2011; Boykoff and Boykoff, 2007). This makes the media a battlefield as those concerned about climate change and their opponents compete to get coverage of their views into the wider public domain and control the way those views are presented. Those concerned about climate change seek to persuade the public that climate change is occurring, is caused by human activity, and is a serious problem that needs to be addressed by government action. Typical strategies involve showing images of expanding deserts, retreating glaciers, and polar bears adrift on ice floes. Their opponents seek either to question whether climate change is happening or persuade that everything is a consequence of natural events that defy human remedy. Typical strategies include giving sceptical scientists public opportunities to air their views and emphasising divisions in the scientific community. The purpose is not simply to counter arguments about climate change but also to confuse the public by stressing uncertainty and disagreement.

Media coverage of climate change in the United States has fluctuated over time (Boykoff, 2011, 25). Studies of major newspaper and national television news coverage of climate change reveal spikes in media interest in the issue that correspond with specific events such as congressional hearings, international summits, film releases, and suggestions of scientific wrongdoing. The percentage of air time and column inches devoted to climate change, however, has remained low. Climate-related stories have constituted less than 1 per cent of all news coverage over the last decade (Boykoff, 2011, 24). Norms of balanced and fair reporting, combined with effective media strategies by groups opposed to action to tackle climate change, have shaped the content of much of this coverage. Studies show that the media has usually provided space to both climate scientists and sceptics in an effort to be seen as fair. This practice is probably even more pronounced in local television news (Boykoff, 2010). The effect is to portray an image of conflict and uncertainty in climate change that is at odds with the consensus proclaimed by scientific bodies. Conservative talk radio, cable news, and syndicated conservative columnists amplify this message of scepticism. Radio host Rush Limbaugh and Fox News presenters Glenn Beck, Bill O'Reilly, and Sean Hannity, for example, have consistently ridiculed the idea of climate change in their shows (Dunlap and McCright, 2011, 153). In recent years these media outlets have been augmented by conservative blogs that promote the message of climate scepticism.

Opinion polls suggest that media coverage of climate change in the United States has shaped public understanding of the issue. Most polls show that a majority of Americans believe that global warming is occurring but the percentage which believes that it is caused by human activity is less convincing and below that found in most other countries. The percentage of Americans who believe that global warming is occurring varies slightly depending on the precise wording of the question asked, but most suggest that a majority believes there is solid evidence of global warming, that it is happening, or already begun (Jones, 2014; Pew, 2009; Borick et al., 2011). These majorities are lower than found in most other developed countries, however, and subject to variation over time (Ray and Pugliese, 2010). Public belief in the existence of global warming, in short, appears

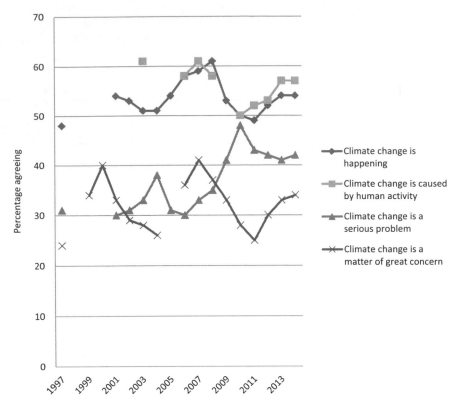

Figure 1.3   Public opinion on climate change, 1997–2014

Source: Based on Gallup Poll data.

susceptible to media campaigns by climate change sceptics. The media coverage
given to climate sceptics also shapes opinion about the scientific consensus on
the issue and the cause of climate change. Americans underestimate the scientific
consensus about climate change. The percentage of Americans who think that
most scientists believe that global warming is occurring has ranged from 48 per
cent to 65 per cent over the last 15 years (Duggan, 2014). Americans are less likely
than their counterparts in other developed countries to blame human activity for
global warming. Between one third and one half of Americans believe that global
warming is caused by natural phenomena (Saad, 2014).

Although a majority of Americans believe that anthropogenic climate
change is occurring, opinion polls suggest that they doubt the seriousness of the
condition. Over the last 15 years the percentage of Americans who believe that
the seriousness of global warming has been exaggerated by the media has ranged
from 31 per cent to 48 per cent, while over the same period the percentage that
regard the reports as generally correct has varied from 23 per cent to 34 per cent

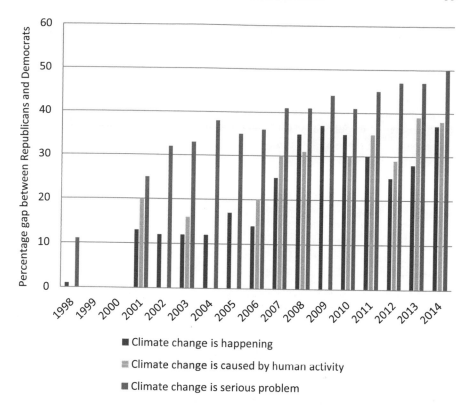

Figure 1.4   Partisan gap in public opinion on climate change, 1998–2014
Source: Based on Gallup Poll data.

(Duggan, 2014). Low levels of concern about climate change reflect this sense that the seriousness of the condition has been exaggerated. Between 1998 and 2014 the percentage of Americans stating that climate change posed "a threat to their way of life" rose slowly from 25 per cent to 36 per cent (Jones, 2014). After more than three decades of warnings about impending catastrophe, nearly two-thirds of Americans did not believe that climate change threatened them in a serious way. This lack of concern is shown in low levels of "worry" about the problem. The percentage of Americans stating that they worry "a great deal" about global warming has varied between 24 per cent and 41 per cent over the last 25 years (Newport, 2014). When ranked against other environmental problems, global warming consistently occupied last place. Public concern is focused on environmental problems which can be experienced directly such as the pollution of drinking water and toxic waste contamination rather than climate change.

Partisanship has emerged as an important feature of public opinion on climate change over the last 25 years. Opinion polls reveal that Democrats and

Republicans have increasingly different views about whether climate change is occurring, its causes, and its seriousness (Dunlap, 2008). In 2014, 73 per cent of Democrats believed that global warming was already happening compared with 36 per cent of Republicans; 79 per cent of Democrats identified human activity as the main cause of climate change compared to 41 per cent of Republicans; and 56 per cent of Democrats worried a great deal about global warming compared to 16 per cent of Republicans (Jones, 2014; Saad, 2014; Newport, 2014). Although these partisan differences in opinion arise from a range of factors, including ideological and cultural values, the *growing* gap in Republican and Democratic attitudes towards climate change owes much to signals from party leaders and conservative media outlets (Brewer, 2012). Since the 1990s Republican and Democratic elites have taken increasingly divergent views on climate change and these have acted as cues that shape public opinion on the issue. The result is that climate change has become one of the issues that defines what it means to be a Democrat or a Republican in the United States (Nisbet, 2011). Some commentators have even suggested that climate change has joined issues such as abortion, same-sex marriage, and school prayer as part of the "culture wars" that have raged in the United States since the 1970s (Kaufman, 2010; Gerson, 2012; Gillis and Kaufman, 2012).

Public opinion on climate change has provided few incentives for policymakers to engage in a major effort to address the problem over the last 25 years. Climate change has lacked the saliency or opinion intensity that is usually regarded as necessary for radical policy change (Nisbet, 2011). A majority of Americans may believe that climate change is occurring, and may even agree with general statements that something should be done to address the problem, but the lack of concern about the issue makes it difficult to mobilise the public to support action. When the costs of proposed solutions become apparent public support tends to evaporate. The emergence of a deep partisan divide has acted as a further barrier against major action. Finding consensus on a solution is difficult when the public is deeply split over the existence of a condition, its causes, and the threat that it poses.

Interest Groups

Interest groups have long been recognised as a central feature of American politics. They offer ways of participating in politics, define problems and offer solutions, mobilise public and elite support for particular policy positions, and persuade or cajole decision-makers to take or block action. To achieve these objectives they engage in activities ranging from media campaigns to meetings with individual decision-makers. How well groups perform these activities undoubtedly shapes policy. Some will have the backing and resources to make their case while others will struggle to be heard amidst a cacophony of competing voices. Groups that are heard have an opportunity to influence how problems and solutions are viewed that is denied to those unable to compete. This is not to argue that policy is *determined* by groups but simply to point out that they play an important *role* in agenda-setting, policy formation, and policy adoption. The constellation of groups active in a policy area matters to policy outcomes.

Two broad competing interest group "camps" have been identified as active in climate change politics (Sussman and Daynes, 2013, 132). The first consists of a range of groups committed to campaigning for government action to address climate change. These include mainstream environmental groups such as the Natural Resources Defense Council, the Sierra Club, and the Environmental Defense Fund, as well as think tanks such as the PEW Research Center, and, in recent years, a variety of groups campaigning on social, welfare, medical, and similar issues. Groups representing the renewable energy industry have also joined this "camp" on occasion. The second consists of a coalition of interests opposed to government action on climate change. These include both industrial groups opposed to the regulation of greenhouse gases on business grounds, and conservative groups concerned about government action on ideological grounds. Prominent among the industrial groups are the fossil fuel industry, the electrical generating industry, and a range of energy-intensive manufacturers. Although important differences have emerged between these groups at various times, they all believe to some extent that government action to tackle climate change would harm their interests. Conservative groups oppose action on climate change out of fear that proposed solutions will lead to greater government regulation and undermine faith in free markets. Think tanks such as the Heritage Foundation and the CATO Institute play a major part in promoting such ideas (Mirowski, 2014; Jacques et al., 2008; McCright and Dunlap, 2003).

The disparity in resources between these two blocks is considerable. Opponents of action on climate change include some of the largest industries in the United States with resources to fund a range of lobbying activities while environmental groups lack such financial muscle. Data compiled by the Center for Responsive Politics gives an indication of the money that key elements in the two blocks have spent on lobbying (see Figure 1.5). Although not all of this expenditure was on matters related to climate change, the amount of money spent by the oil and gas industry and electrical utility companies has dwarfed that spent by environmental groups. Even when the spending of the alternative energy industry is added to that of environmental groups the amount remains a fraction of that spent by the oil and gas industry and electrical utility companies. A similar picture emerges when campaign contributions to political parties and candidates are examined. The Center for Responsive Politics data shows that the oil and gas industry and electrical utility companies have made much larger contributions to election campaigns than environmental groups and the alternative energy industry (see Figure 1.6). A distinct partisan bias has been apparent in these contributions with groups opposed to action on climate change giving overwhelmingly to Republicans and those in favour of action giving to Democrats. In the 2012 election cycle 80 per cent of the contributions made by the oil and gas industry went to Republicans and 90 per cent of the contributions made by environmental groups went to Democrats.

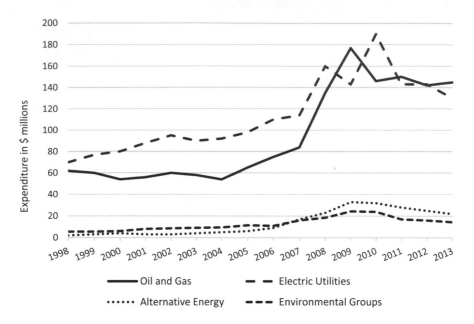

Figure 1.5   Expenditure on lobbying by selected sectors, 1998–2013
Source: Based on data compiled by OpenSecrets.org.

Groups opposed to government action on climate change have used three main tactics to achieve their goals (Layzer, 2007). First, they have engaged in traditional "insider" methods to present their arguments directly to decision-makers and their staff. Access has been facilitated by their sizeable campaign contributions, professional lobbyists have been hired in their hundreds to make a case, and the importance of fossil fuels in the national and local economies has ensured that their voices have been heard and noted. Second, groups have sought to persuade the public that climate change is not a problem through the "manufacturing" of uncertainty and doubt (Dunlap and McCright, 2011, 146; Oreskes and Conway, 2010; Begley, 2007; Gelbspan, 2004). Front groups such as the Global Climate Coalition, the Information Council on the Environment, and the Cooler Heads Coalition, together with conservative think tanks such as the Heritage Foundation and the CATO Institute, have engaged in public relations campaigns that promote climate change scepticism (McCright and Dunlap, 2003). Another key element in this battle to influence public opinion has been to stress the economic costs that would result from action to address climate change. Front groups and conservative think tanks have consistently argued that action would lead to job losses and loss of economic competitiveness. Third, groups have tried to create an impression of "grassroots" opposition to action on climate change by creating "astroturf" groups (Dunlap and McCright, 2011, Pooley, 2010). The American Petroleum Institute, for example, created an "astroturf" group called

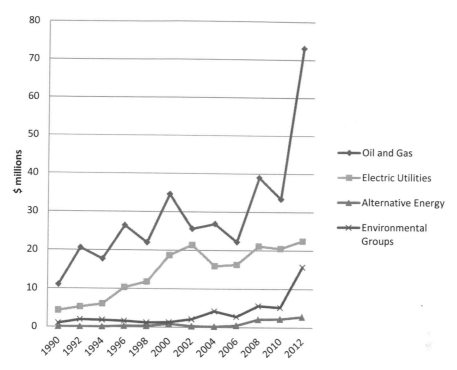

Figure 1.6   Federal campaign contributions by selected sectors, 1990–2012
Source: Based on data compiled by OpenSecrets.org.

Energy Citizens to campaign against climate change legislation in the early years of the Obama Administration.

Groups advocating action to tackle climate change have often struggled to match the persuasive power of their opponents. This is not simply due to the mismatch in resources available to the two sides, but also a consequence of what they are trying to sell. Making a case for a change in policy is nearly always more difficult than arguing to maintain the status quo (Riker, 1996, 69; Jerit, 2008). Proponents of change must explain the proposed change *and* defend it from attack, whereas supporters of the status quo need only highlight the shortcomings of any proposals for change. In the case of climate change these difficulties are compounded by the evidentiary basis of the condition, uncertainty about its consequences, and a lack of vision about the future. Environmental groups need to persuade decision-makers and the public that climate change is occurring and requires urgent attention despite a lack of direct immediate evidence to support such claims. To do this they have typically engaged in media campaigns that stress impending catastrophe and crisis which incorporate images of retreating glaciers, floods, and violent storms. Doubt has been cast, however, on the effectiveness of media campaigns based on fear appeals (Moser and Dilling, 2011). A focus on negative images without a discussion of feasible solutions usually leads to denial rather than mobilisation. Environmental groups have also been criticised for concentrating on what they oppose rather than articulating a vision of the sort of future they want (Shellenberger and Nordhaus, 2004; Bryner, 2008). The charge is that environmental groups have failed to explain what a post-fossil fuel economy would look like and how this fits with dominant American values.

The balance of interest groups involved in climate change policy has presented substantial obstacles to proponents of government action over the last forty years or so. Groups opposed to action have had substantial resources and been able to exploit suggestions of scientific uncertainty and economic costs, whereas groups favouring action have had fewer resources and struggled to sell their message effectively. This is not to argue that groups advocating action on climate change have failed to secure any of their policy goals, or that groups opposed to action have been successful at all times. Both sides have had successes and failures. The essential point is that the balance of interest groups has helped to frustrate efforts to produce a major change in policy—not that it has stopped all policy change. Adjustments in policy have certainly been shaped by interest group activity at particular times over the last 30 years (Bryner, 2008).

Control of Political Institutions

The control of political institutions is a key element of the "political stream." Political actors such as the president, executive officials, legislators and judges have an impact on how problems and solutions are perceived, the access given to interest groups, and the decision-making process. Changes in these personnel, therefore, may have a powerful impact on the policy process (Kingdon, 2011). Sometimes such changes may make action on a particular issue more unlikely, but at other times they may offer opportunities to consider addressing problems that had previously been ignored. Changes in personnel may not always be sufficient to bring about policy change but they will often raise challenges to the status quo. New people with different ideological orientations and political needs from their predecessors will often re-evaluate priorities, favour different voices in arguments, explore whether alternative venues can be used to further policy goals, and employ a different calculus of costs and benefits when making decisions.

The president is the most important political actor in the policymaking process. A combination of constitutional, delegated, and other powers means that the president is able to command attention, identify government priorities, suggest ways of addressing problems, and initiate certain forms of action. Presidents can use the annual State of the Union address and other major speeches to bring attention to problems and suggest solutions. They have the authority to negotiate with foreign leaders. They can recommend legislation to Congress and veto bills they do not favour. They can issue executive orders requiring federal bureaucracies and agencies to take action. And they can use the regulatory rule-making process to change policy in certain circumstances. Powers of appointment to top level posts in the Administration and a large number of advisory and other bodies means that they can also influence the range of voices that are heard in the corridors of power. The ability to make appointments to the federal judiciary can offer further opportunities to shape an important element of the policymaking process. These powers do not provide the president with the necessary authority to achieve their policy objectives in all cases because of the system of checks and balances built into the American political system, but they certainly place the occupant of the White House at the centre of the policymaking process.

Control of the White House has been equally shared between Democrats and Republicans since climate change emerged as a political issue in the late 1970s. These changes in partisan control have produced considerable shifts in presidential leadership on climate change, with Jimmy Carter (1977–1981), Bill Clinton (1993–2001), and Barack Obama (2009–2017) generally supportive of action to address the problem and Ronald Reagan (1981–1989), George H.W. Bush (1989–1993), and George W. Bush (2001–2009) either lukewarm about the need to do anything or actively opposed to government action. These differences in attitude matter. Presidential support for action on climate change can provide important leadership in both the international and domestic arenas, and can lead to significant changes in regulatory approaches to the problem while opposition can undermine international efforts to negotiate treaties, make successful legislative action unlikely, and can lead to reversals in policy through administrative means (Klyza and Sousa, 2008; Shafire, 2014). Achieving policy objectives is conditional, however, even for presidents. Other pressing concerns such as the economy or war can divert attention away from climate change, Congress may oppose or seek to reverse presidential actions, and public support for action may be difficult to mobilise. The president has considerable power but cannot command the political system to do his will.

Presidents have often experienced considerable difficulty getting legislative support for their policy proposals. Congress is an independent legislature possessing a range of important powers with a membership capable, and frequently willing, to challenge presidential initiatives or advance proposals of their own. Congress can enact laws, has budget-making authority, can reject treaties negotiated by the president, can turn down presidential appointments to the executive branch and the judiciary, and has important oversight and investigatory powers. This array of powers gives Congress the potential to play a major role in the policymaking process. A number of structural obstacles, however, need to be overcome if that role is to be performed to the full. Bicameralism acts as a brake on law-making as the different makeup of the House of Representatives and the Senate can make reaching the agreement needed to enact legislation very difficult. The committee system can also frustrate law-making as complex proposals (like most of those dealing with climate change) usually get referred to multiple committees for consideration which creates problems of co-ordination and opportunities for obstruction. Further opportunities for obstruction exist in the Senate where the right of senators to filibuster legislation makes it difficult to end debate and vote on a measure.

Political parties can either smooth over many of these structural obstacles to policymaking or turn them into insurmountable barriers. If a political party controls both the House and the Senate with large unified majorities, or if a strong element of bipartisanship exists on an issue, the problems posed by bicameralism, a fragmented committee system, and the occasional maverick senator can usually be overcome. If different political parties control the House and the Senate, if majorities are small and unstable, or if the parties are polarised on an issue, the prospects of policymaking become less. Numbers, unity, and policy orientation are the key factors here. Which party controls which chamber? Are the parties unified or divided on an issue? Are there significant differences between the parties on the issue?

Table 1.1   Party control of Congress and the presidency

	Party Control of House of Representatives (and Size of Majority)	Party Control of Senate (and Size of Majority)	Party Control of the Presidency
95th Congress (1977–1978)	D (149)	D (23)	D
96th Congress (1979–1980)	D (119)	D (17)	D
97th Congress (1981–1982)	D (50)	R (7)	R
98th Congress (1983–1984)	D (103)	R (8)	R
99th Congress (1985–1986)	D (71)	R (6)	R
100th Congress (1987–1988)	D (81)	D (10)	R
101st Congress (1989–1990)	D (85)	D (10)	R
102nd Congress (1991–1992)	D (100)	D (12)	R
103rd Congress (1993–1994)	D (82)	D (14)	D
104th Congress (1995–1996)	R (26)	R (4)	D
105th Congress (1997–1998)	R (19)	R (10)	D
106th Congress (1999–2000)	R (12)	R (10)	D
107th Congress (2001–2002)	R (9)	D/R (1)	R
108th Congress (2003–2004)	R (24)	R (2)	R
109th Congress (2005–2006)	R (29)	R (10)	R
110th Congress (2007–2008)	D (37)	D (2)	R
111th Congress (2009–2010)	D (79)	D (16)	D
112th Congress (2011–2012)	R (49)	D (6)	D
113th Congress (2013–2014)	R (33)	D (10)	D
114th Congress (2015–2016)	R (59)	R (10)	D

The answers to these questions have created problems for advocates of government action to address climate change. Since the late 1970s periods of unified government have been relatively rare (see Table 1.1). Democrats have controlled the presidency and Congress for eight years while Republicans have controlled these institutions for six years. Unified control of Congress has been slightly less rare. Democrats have controlled both the House and the Senate for 16 years while the Republicans have controlled both chambers for 14 years. Divided government does not necessarily lead to legislative gridlock. Mayhew (1991) has shown that major legislative action is possible under divided government. Changes in the composition of the congressional parties over the last four decades, however, have served to increase the likelihood of stalemate. The parties have become more ideologically polarised and unified. This is particularly the case on environmental issues in general and climate change in particular. The gap between Democrat and Republican voting records on environmental issues has steadily increased since the late 1970s (see Figure 1.7). Democrats are much more likely to support initiatives to tackle environmental problems like climate change than Republicans. This does not mean that all Democrats favour action on climate change and all Republicans oppose doing anything. Democrats from energy producing states and constituencies often express concern about action that would damage local economic interests, and a few Republicans have been at the forefront of developing proposals to address the problem (Bomberg and Super, 2009). What the voting records suggest is a general disposition of attitudes towards climate change that has shaped congressional action.

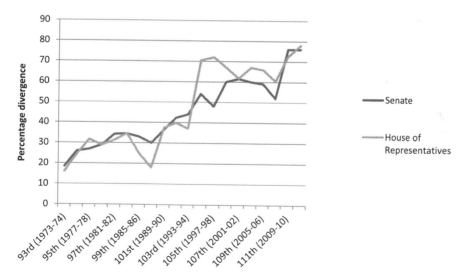

Figure 1.7   **Divergence between Democrats and Republicans on voting on environmental issues**

Source: Compiled from LCV data.

The differences between congressional Democrats and Republicans over climate change has two important consequences. First, majority status provides a party with leadership positions which facilitates agenda setting and the consideration of legislation. Democrats are more likely to use the institutional authority provided by majority status to schedule action on climate change and view proposals more sympathetically than Republicans (Liu et al., 2011; Fisher et al., 2013). Second, defections from party positions have become rare. This makes passing legislation difficult as neither party has had a majority large enough to invoke cloture (end a filibuster) in the Senate since the 95th Congress (1977–1978). The result of these two features has been the creation of a political environment where talk is easy but action difficult (Rabe, 2010b; Sussman and Dugan, 2013). The number of bill introductions and committee hearings addressing climate change has been high at times but subsequent legislative action has been few and far between. Overcoming the barriers that restrict action on climate change has proved difficult.

The role of the Supreme Court and other federal courts in shaping climate change policy has grown over the last three decades as both proponents and opponents of action have pursued litigation to achieve their objectives (Brinkman and Garren, 2011; Engel, 2010). On the one hand, proponents have filed lawsuits that either seek to compel federal agencies to tackle climate change or else hold large industrial emitters liable for their greenhouse gas emissions under the common law. On the other hand, opponents have sought to challenge the legality or scientific basis of regulatory efforts to address climate change. Court rulings in these cases obviously have a direct impact on policy by compelling or stopping regulatory activity, or changing the behaviour of polluters through establishing liability. The policy impact of the Supreme Court and lower federal courts in policymaking, however, can often be broader than establishing a winner and a loser in a particular case. Litigation can play a role in framing a problem and generating policy relevant information (Engel, 2010). Media coverage of court cases may also help shape public opinion about the issue.

Constitutional considerations and statutory interpretation dominate the decision-making of the Supreme Court and lower federal courts. Although attention to precedent and other legal niceties is an important feature of this process, the personal opinions and ideological orientation of the judges play a significant role in determining outcomes. Liberal judges may have an expansive view of the role of the courts and interpret constitutional clauses in one way while conservative judges may prefer a more limited role for the courts and have very different views about the meaning of the Constitution. This means that the composition of the courts is very important. The Constitution gives the president power to appoint Supreme Court justices and other federal judges with the agreement of the Senate, but appointments can only be made when there is a vacancy. An element of luck, therefore, determines a president's ability to shape the composition of these courts. Over the last three decades Republican presidents have had more opportunities to make judicial appointments than their Democratic counterparts, giving the

Supreme Court and many other lower courts a conservative bias. Concerns about the proper scope of the courts in addressing a highly charged and political issue, the limits of the federal government's regulatory authority, and the need to protect public property have often been raised as a result when the courts have considered climate change cases.

Occasions when proponents of action to address climate change have had clear control of key political institutions have been extremely rare during the last 40 years. The first two years of the Obama Administration is the only time when a president strongly committed to taking action enjoyed Democratic majorities in Congress, but even then Republicans had sufficient votes in the Senate to obstruct major initiatives. At other times the political arithmetic needed for major policy change has been even worse. Presidents have either been actively opposed or unsupportive of action, or else faced a hostile Congress. This does not mean that no policy change has been possible. Presidents can often bypass obstructive legislators by taking executive action, winning legislative coalitions can often be negotiated for incremental changes, and court decisions can also lead to changes of policy. The pattern of control of political institutions, however, raises substantial barriers to radical changes of policy.

Policy Windows?

Kingdon (2011) argues that the opportunity for major policy change arises when the problem, policy, and politics streams are brought together to create a "policy window." "The separate streams come together at critical times," he states, "a problem is recognized, a solution is developed and available in the policy community, a political change makes it the right time for policy change, and potential constraints are not severe" (2011, 165). An event or change in any of the streams can provide the catalyst for a "policy window" to open. A crisis or catastrophe might focus attention on a problem that had previously been regarded as a low priority, the development of a new solution might make dealing with a problem more palatable, or an election might bring into office decision-makers with different values and needs. If such a development prompts a "coupling" that joins the streams together a "choice opportunity" is created that offers the possibility of major policy change depending on "the mix of elements present and how the elements are coupled" (2011, 166). In other words "policy windows" create opportunities not outcomes.

"Policy entrepreneurs" play a key role in creating and exploiting opportunities for policy action. Political actors willing to commit resources to advance particular policy action not only campaign to "push" for their proposals but also "lie in wait—for a window to open" (Kingdon, 2011, 181). When "policy entrepreneurs" see a "window" opening they move rapidly to raise awareness of a problem, promote a solution, and negotiate an outcome. In other words, they act as both "advocates" and "brokers." Kingdon argues that this rush by "policy entrepreneurs" to exploit an opportunity is a major factor in bringing the streams together. If a "policy entrepreneur" is not present, "the linking of the streams may not take place" (Kingdon, 2011, 182).

Central to Kingdon's analysis is the idea that open "policy windows" are infrequent and close very quickly. This is particularly the case with climate change, where the characteristics of the three streams mitigate against easy "coupling." The problem has not been regarded as pressing, solutions do not fare well against the "criteria for survivability," and political forces have usually been hostile to radical action. Subsequent chapters will confirm that open "policy windows" have been rare and have closed fairly quickly. This does not mean that no policy change has occurred over the last 40 years. The US government has taken action to address climate change but that action has been incremental rather than radical. US policy on climate change has been characterised by initiatives to sponsor research, promote the development of new technologies, and build upon existing statutory authority to address the problem rather than by major legislative enactments that seek to control greenhouse gas emissions.

Chapter 2
Small Steps to Rio

A study of US climate change policy has a number of possible starting points. The Office of Naval Research and the US Air Force funded basic research into weather prediction and manipulation during the Cold War that allowed scientists such as Gilbert Plass, Roger Revelle, Hans Suess, and Charles (Dave) Keeling to make important theoretical and empirical advances in the scientific understanding of anthropogenic climate change (Hart and Victor, 1993; Fleming, 1998; Weart, 2008). Roger Revelle made the first public call for federal funding of research into global warming, which Congress subsequently approved, at a congressional committee hearing in 1956 (Keller, 2009). President Lyndon Johnson noted in a Special Message to Congress on 8 February 1965 that: "This generation has altered the composition of the atmosphere on a global scale through ... a steady increase in carbon dioxide from the burning of fossil fuels." None of these possible starting points, however, really marks the beginnings of the US government's policy towards climate change, as the primary focus of all was something else. Better candidates for the start of US policy occur in the late 1970s when policymakers began to consider anthropogenic climate change more directly. In 1976 a congressional committee held a hearing on legislation introduced by Rep. Phillip H. Hayes (D. IN) that required the federal government to conduct research on the influence of human activities on the climate, among other things. The following year President Jimmy Carter asked the Council on Environmental Quality and the Department of State to prepare a study of global environmental issues, including climate change, that culminated in the eventual publication of the *Global 2000 Report to the President* in 1980.

In the decade or so that followed these first stirrings of policy interest in anthropogenic climate change, the United States government took a number of small steps to address the issue. Laws authorising research were enacted in 1978, 1987, and 1990; the Energy Policy Act of 1992 included provisions to improve energy efficiency; the Bush Administration launched programmes to plant a billion trees to help absorb atmospheric carbon dioxide and persuade large businesses to switch to energy efficient lights; and the United States engaged (albeit often unenthusiastically) with international efforts to address the issue which culminated in the United Nations Framework Convention on Climate Change (UNFCCC) of 1992. This pattern of small policy steps reflected the interplay of the problem, policy, and politics streams at the time. First, questions about whether a condition existed and constituted a problem featured widely in debates during this period despite an emerging consensus among scientists that anthropogenic climate change was occurring and would have adverse consequences. Economic problems, national security concerns, and even environmental problems such as clean air and water appeared far more pressing to the majority of the public and most decision makers. Second, policy knowledge was limited with no agreement about the best way to deal with the problem. Scientists may have grown more confident about the science of climate change during this period but they were far less confident about advocating solutions. Finally, the political environment during these formative years acted as a barrier against radical policy action. A neoliberal sentiment hostile to government regulation dominated the public mood, Republicans controlled the presidency for most of the period, and the energy industry mobilised earlier and more effectively than environmental groups to influence policy responses.

The small policy steps taken by the US government to address climate change had little immediate impact upon greenhouse gas emissions, but that does not mean they were unimportant. The research effort set in motion during this period not only provided improved knowledge about the problem, but also created institutional settings where scientific findings could be discussed and publicised. The international processes created to address climate change during this period institutionalised concern still further and helped to keep the issue on the agenda. Both the Intergovernmental Panel on Climate Change (IPCC), created in November 1988, and the Conference of the Parties (CoP) meetings that followed ratification of the UNFCCC in 1992 established timetables, venues, and pressure for continued engagement with the issue that would prove difficult to ignore completely.

The Beginnings of US Climate Change Policy

Funding by the US military played an important role in the development of climate science following the Second World War. Driven by the imperatives of the Cold War this funding enabled Roger Revelle and Hans Suess to improve understanding of the way that oceans absorbed atmospheric carbon dioxide, gave Gilbert Plass the opportunity to study how atmospheric carbon dioxide absorbed infrared radiation, and provided Dave Keeling with resources to measure the amount of carbon dioxide in the atmosphere (Hart and Victor, 1993; Fleming, 1998; Weart, 2008). Keeling's observations provided the first empirical proof that carbon dioxide levels in the atmosphere were rising year by year. Dating the start of US government policy on climate change to this funding, however, would be a mistake. Although the theoretical and empirical advances that flowed from spending by the military during this period undoubtedly laid many of the foundations of modern climate science, there is no evidence that this was the purpose of the funding. The military were concerned with improved weather prediction, greater understanding of ocean currents, and the possibility of manipulating the weather for strategic advantage rather than investigating potential causes and consequences of climate change (Weart, 2008, 22; Kwa, 2001).

The first public request for federal funding to investigate climate change occurred in 1956 when Roger Revelle gave evidence before a congressional committee looking at funding for the International Geophysical Year of 1957–1958 (Keller, 2009, 61). Revelle told the House of Representatives' Appropriations Committee that global warming was a "tremendous geophysical experiment" that warranted research, and compared his request for funding to Christopher Columbus asking "the sovereigns of Spain" to finance his voyages of discovery (HCoA, 1956, 465–7). Congress subsequently approved limited funding that helped resource Dave Keeling's work; but again, this action does not mark the beginning of US climate change policy. The decision to provide funds for the International Geophysical Year was driven primarily by Cold War imperatives. The US government hoped that support for the International Geophysical Year would enhance national prestige and generate data that had military value (Weart, 2008, 33). Support for Revelle's work was a small by-product of this larger purpose and not the start of government concern about the possibility and consequences of anthropogenic climate change.

The results of this early research into climate change began to attract the attention of federal government scientific bodies in the 1960s. In 1965 the President's Science Advisory Committee created a panel to review environmental concerns that included a subpanel on climate change. Their report concluded that: "By the year 2000 the increase in atmospheric CO_2 ... may be sufficient to produce measurable and perhaps marked changes in climate ... that could be deleterious from the point of view of human beings" (PSAC, 1965, 126–7). President Lyndon Johnson referred to the report in a Special Message to Congress on Conservation and the Restoration of Natural Beauty delivered on 8 February 1965 when he noted: "This generation has altered the composition of the atmosphere on a global scale through ... a steady increase in carbon dioxide from the burning of fossil fuels." Johnson's statement, however, revealed a concern about air pollution, not climate change. He compared the build-up of carbon dioxide in the atmosphere with the increase in radioactivity caused by nuclear testing. No scientific consensus existed at the time about whether a condition existed or whether it was a problem. A report by the National Academy of Sciences (NAS, 1966) concluded that climate change was possible but posed no immediate cause for concern. The NAS recommended further research.

Federal funding of basic research on climatic systems continued on a piecemeal basis in the late 1960s and early 1970s as a by-product of other projects. The US military began to collect meteorological data from spy satellites; scientists such as James E. Hansen at NASA's Goddard Institute for Space Studies started to apply weather models developed to understand the atmospheres of other planets to the Earth; and the National Oceanic and Atmospheric Administration (NOAA), created in 1970, provided funding for the study of the ocean (Weart, 2008, 90–91, 95–6). This funding contributed to a growing scientific literature about the link between carbon dioxide in the atmosphere and global warming, had started to identify feedback mechanisms that suggested that climatic change could occur rapidly rather than over centuries, and could point to potential catastrophic consequences if nothing was done. An NAS report on "Energy and Climate" concluded in 1977 that climate change caused by human activity might lead to rises in sea levels, disruption to fisheries, crop failures, and drought, but stressed that scientists could not foresee what would actually happen on a particular timescale (NAS, 1977).

Droughts in Africa and Asia leading to widespread famines, and crop failures in the American Midwest, appeared to give credence to scientific warnings about climate change and sparked increased media interest in the subject during the mid-1970s (Weart, 2008, 87). Headlines such as "Ominous Changes in the World's Weather" began to appear in newspapers and journals with stories talking variously about the possibility of a new Ice Age or a period of warming (Alexander, 1974). This media coverage eventually prompted congressional interest in the climate. In 1976 Rep. Phillip H. Hayes (D. IN) introduced "The National Climate Program Act" that included provision for research into the "influence of man's activities on the process of climate dynamics." The House Committee on Science, Space and Technology's Subcommittee on the Environment and Atmosphere held a hearing on the bill in May 1976, but no legislative action followed (HCSST, 1976). Rep. George E. Brown (D. CA) introduced a new version of the bill the following year, and further hearings were held by committees in both the House and Senate. Although the testimony at these hearings revealed a lack of consensus among scientists about the reasons why government should fund further research into climate change, with some identifying a potential threat to food production and others a possible risk to water supplies, all agreed that further study was needed (Keller, 2009, 102). Congress subsequently enacted the National Climate Act of 1978 which established a national climate research programme that included a mandate for the study of anthropogenic climate change. A national Climate Program Office was established within NOAA to administer the programme (Pielke, 2000).

These stirrings of congressional interest in climate change in the late 1970s were mirrored in the White House. President Jimmy Carter made an important contribution to the development of climate change policy early in his presidency when he signalled a concern about global environmental problems. In "The Environmental Message to Congress" on 23 May 1977 he noted that: "Environmental problems do not stop at national boundaries" and "we and other nations have to recognize the urgency of international efforts to protect our common environment," before announcing that he had directed various government agencies "to make a one-year study of the probable changes in the world's population, natural resources, and environment through the end of the century." No explicit mention of climate change was made in "The Environmental Message to Congress," but Carter's concern about the global environment made him increasingly appreciate the potential threat posed by the problem (Sussman and Daynes, 2013, 80). In his "Science and Technology Message to the Congress" on 27 March 1979 he became the first president to mention anthropogenic climate change in a public speech. In this "Message" he stated that: "Another problem we face is the risk that man's own activities—now significant on a global scale—

might adversely affect the earth's environment and ecosystem. Destruction of the ozone layer, increase in atmospheric carbon dioxide, and alteration of oceanic flow patterns are examples." Further recognition that climate change was a problem that needed addressing came with the publication of a report detailing the findings of the study that President Carter had initiated in 1977. The *Global 2000 Report to the President*, published in July 1980, devoted a chapter to climate change. The report stated that "some human activities, especially those resulting in releases of carbon dioxide into the atmosphere, are known to have the potential to affect the world's climate" and cautioned that "many experts … feel that changes on a scale likely to affect the environment and the economy are not only possible but probable in the next 25 to 50 years" (cited in Sussman and Daynes, 2013, 80).

Enactment of the National Climate Act of 1978 and President Carter's actions that led to the publication of the *Global 2000 Report to the President* mark the tentative beginnings of US climate change policy. They reveal awareness that a condition might exist with possible adverse consequences, and establish a policy response *intended* to find out more about the problem. These apparently limited actions should not be underestimated. The requirement for federal agencies to conduct further research into climate change ensured that it would be difficult for the issue to disappear completely from the political agenda as organisations such as the Environmental Protection Agency (EPA), NOAA, and NASA released periodic reports on their findings. Publication of a report by the NAS on *Carbon Dioxide and Climate* in 1979, for example, prompted the Senate Committee on Energy and Natural Resources to hold the first congressional hearing devoted solely to climate change in April 1980 (Keller, 2009, 103). In Kingdon's terminology these reports provide "systematic indicators" that a problem might exist (Kingdon, 2011, 90). Unsurprisingly the beginnings of US climate policy also made clear some of the obstacles that would shape policy development in the future. First, a concern about energy supplies meant that President Carter supported initiatives to increase the use of coal at the same time as warning about the build-up of greenhouse gases in the atmosphere. Demands for abundant and cheap energy compromised action to address climate change from the beginning. Second, the fossil-fuel industry mobilised quickly to deny that there was any need for action. Forging alliances with conservative think tanks and media outlets they sought to persuade policymakers and the public that climate change was nothing to worry about (Weart, 2008, 110).

The beginnings of US climate change policy can be traced to changes in the problem, policy, and politics streams in the late 1970s (Kingdon, 2011). By this time sufficient scientific evidence had been amassed to suggest that a condition existed that might pose some sort of threat to the country's well-being, extreme weather events appeared to give credence to the arguments of climate scientists and generated public concern, politicians had been elected to the White House and Congress who were sympathetic to such concerns, and few disputed that increased research offered the best response. Streams are not stable, however, and the opportunity for significant further action faded quickly. Although scientists continued to provide evidence about the causes and consequences of climate change other concerns seemed more immediate and important, possible solutions appeared costly, and the election of President Reagan and a Republican take-over of the Senate in 1980 presaged an anti-regulatory mood in Washington, DC, that regarded environmental initiatives with suspicion if not outright hostility.

From Indifference to Renewed Concern

The 1980 elections produced important changes in political stream. President Reagan entered the White House with a promise to "take government off the backs of the great people of this country," and Republicans gained a majority in the Senate for the first time since 1954. These electoral triumphs heralded the "Reagan Revolution" in which environmental concerns had little place in a quest to reduce the size of government and cut taxes (Evans and Novak, 1981; Kymlicka and Matthews, 1988). EPA Administrator Anne Burford and Interior Secretary James Watt obstructed the implementation of laws, reduced budgets, and sided with business interests in disputes over public lands, mining, waste disposal, and a range of other environmental issues (Shanley, 1992). The administration showed little interest in climate change, and proposed dramatic cuts in research funding in the early 1980s, though seldom secured congressional approval for these reductions in their entirety. Climate change never completely disappeared from the political agenda, however, as US government agencies continued to circulate the findings of their research, international organisations began to focus on the issue, and congressional proponents of action kept the issue alive by holding hearings. The cumulative effect of these pressures eventually led to increased prominence being given to climate change in the last years of the Reagan Administration leading to enactment of the Global Climate Protection Act of 1987. A record heat wave in 1988, dramatic testimony by NASA scientist James Hansen before a Senate committee, and rising public concern about "the greenhouse effect" would eventually propel the issue into the presidential election of 1988.

The publication of reports on research mandated by the National Climate Act of 1978 ensured that climate change remained on the public agenda despite the indifference of the administration on the issue. In 1983 reports by the NAS and EPA published within three days of each other offered similar assessments of the science of climate change but drew different conclusions about the consequences and need for action (Weart, 2008, 141). The contribution of economists to the NAS report explains the differences between the two publications as they emphasised the cost of mitigation compared to adaptation (Oreskes and Conway, 2010, 177–82). Both confirmed that the evidence suggested that increasing levels of carbon dioxide in the atmosphere was contributing to climate change, but while the NAS report concluded that any warming would not be too severe and could probably be accommodated with little disruption, the EPA report warned that temperature rises would be large with potentially catastrophic consequences felt within a few decades (Weart, 2008, 141). This was the first time a government agency had stated that climate change posed a real threat rather than a theoretical one though its impact was undermined by the NAS report. When asked to comment the administration used the NAS report to discredit the EPA report (Oreskes and Conway, 2010, 182). Other reports followed, however, that reinforced the EPA's warnings. In 1986 a NASA report claimed that the increase of greenhouse gases in the atmosphere was a "totally uncontrolled experiment with no kind of knowledge of where we are going in the end" (Keller, 2009, 69).

International voices began to augment domestic claims that a condition existed that posed a potential threat during the 1980s. Climate change first emerged as a significant issue on the international *scientific* agenda in 1979 when the World Meteorological Organisation (WMO) and the United Nations Environment Program (UNEP) held a World Climate Conference in Geneva (Cass, 2006, 21; Bolin, 2007, 31). Participants at the conference established the World Climate Program (WCP) to further climate research and publicise findings. A year later the WCP organised a conference at Villach, Austria, which concluded that the increase in atmospheric concentrations of greenhouse gases posed a threat to the equilibrium of the earth's climate and needed to be addressed (Morrissey, 2001, 54; Torrance, 2006, 43). Subsequent meetings at Villach revealed a growing consensus among scientists from around the world about the reality and dangers of anthropogenic climate change. In a significant development the 1985 Villach conference reported that knowledge of the potential threat posed by climate change had reached a point that warranted a policy response and called for scientists and policymakers to work together "to explore the effectiveness of alternative policies and adjustments" (Bodansky, 1993, 460). UNEP responded by making the negotiation of an international agreement on climate change one of its long-term goals (Cass, 2007, 21).

The emergence of climate change as an international *political* issue generated pressure on the US government to take action in the mid-1980s. Following the 1985 Villach conference, the executive director of UNEP wrote to US Secretary of State George Shultz to suggest that the United States organise a response to the threat of climate change (Cass, 2007, 21). Discussions about the best way forward began in 1986 and eventually led to American support for a new international process to assess the science, consequences, and policy implications of climate change (Agrawala, 1998; Bolin, 2007, 47). At a Congress of the WMO held in May 1987 the United States proposed that the WMO and UNEP should create a new panel to study the scientific issues surrounding climate change. This finally led to the creation of the Intergovernmental Panel on Climate Change (IPCC) in November 1988. The administration's decision to support the creation of the IPCC reflected a desire to postpone action to address climate change out of concern that such action would undoubtedly harm American industry (Hecht and Tirpak, 1995). Like earlier domestic initiatives to promote research, however, the creation of the IPCC institutionalised concern about climate change and helped keep the issue on the political agenda. Subsequent IPCC reports would be used to highlight the issue and urge a response from the US government.

An indication of the way in which official research findings act as "systematic indicators" can be seen in the response of Congress to the publication of scientific reports by domestic and overseas agencies during the 1980s. These reports provided opportunities for congressional proponents of action on climate change to hold hearings that explored and publicised the issue. During the early 1980s Rep. Al Gore (D.TN) persuaded the House Committee on Science, Space and Technology to hold hearings on a number of reports about rising levels of atmospheric carbon dioxide (Pielke, 2000). The Senate Committee on Environment and Public Work's Subcommittee on Toxic Substances and Environmental Oversight discussed the findings of the 1985 Villach conference report in a hearing in December 1985, and the full committee covered the NASA report in a hearing in June 1986 (SCEPW, 1985; SCEPW, 1986). These hearings revealed a growing divergence between the testimony of scientists and administration officials. Scientists generally warned of the threat of climate change while government witnesses usually played down the dangers. In effect, the administration adopted a "wait and see" approach towards climate change (Cass, 2006, 32). Officials argued that more research was needed to understand fully the problem and potential consequences before deciding what to do. They suggested that action before all the facts were known would be premature. Testimony from witnesses apart from those representing the Reagan Administration overwhelmingly supported more immediate action to address climate change (Keller, 2009, 105).

The testimony given at the hearings of the mid-1980s prompted further congressional interest in climate change. Victories in the 1986 mid-term elections gave Democrats control of the Senate as well as the House and opened opportunities for legislative action. In 1987 seven hearings were held on climate change which matched the number for the previous six years (Keeler, 2009, 104). The momentum generated by these hearings led to the enactment of the Global Climate Protection Act of 1987 (formally a title of the Foreign Relations Authorization Act, Fiscal Years 1988 and 1989) in December 1987. The law gave responsibility for the development of climate change policy to the EPA and State Department, and required annual reports to Congress on the science and implications of climate change. President Reagan signed the law but opposed congressional efforts to transfer control over policy development from the White House Office of Science and Technology Policy to other parts of the executive branch (Pielke, 2000). Reagan preferred to maintain control of climate change policy within the White House where discussions could be better shielded from congressional investigations. The administration's "wait and see" approach to climate change meant that it was happy to agree additional funding for research but was wary of losing control of policy development. Funding for climate research increased substantially during the last years of the Reagan Administration despite reductions in other environmental programmes (Cass, 2006, 32)

An exceptionally hot summer in 1988 meant that congressional interest in climate change continued following enactment of the Global Climate Protection Act of 1987. Media reports of heat waves, droughts, and forest fires prompted further congressional hearings on the topic. Eight committees held 10 hearings on climate change in 1988 as legislators scrambled to respond to the concerns of their constituents about the unusual weather (Keeler, 2009, 69). The most dramatic moment in these hearings occurred when NASA scientist James Hansen told the Senate Committee on Energy and Natural Resources that he was 99 per cent certain that a long-term warming trend was underway that could not be attributed to natural factors (SCENR, 1988). Extensive media coverage of Hansen's testimony followed with *The New York Times* putting the story on its front page (Shabecoff, 1988). Not all scientists agreed with Hansen's bold claim, but the media attention generated propelled climate change into a presidential election for the first time. Both party platforms mentioned the issue. The Democratic Party Platform published on 18 July 1988 stated that "regular world environmental summits should be convened by the United States to address ... the 'greenhouse effect'," while the Republican Party Platform published on 16 August 1988 mentioned that a need to "develop international agreements to solve complex global problems such as tropical forest destruction, ocean dumping, climate change ... will be vital in the years ahead." A brief exchange about climate change occurred in the Vice Presidential Debate held in Omaha, Nebraska on 5 October 1988 with both Senator Lloyd Bentsen (D. TX) and Senator Dan Quayle (R. IN) stressing the need to develop alternative sources of fuel. And Vice President George H.W. Bush declared in a campaign speech delivered on 31 August 1988 that: "Those who think we are powerless to do anything about the 'greenhouse effect' are forgetting about the 'White House effect'" (Hecht and Tirpak, 1995, 383). He promised to hold an international conference on climate change within a year of taking office.

Mention of climate change in the 1988 presidential election provides evidence of the changes that occurred in the policy environment surrounding the issue during the Reagan years. First, scientific advances provided increased evidence that a condition existed which might have adverse consequences in the near future. Sufficient uncertainties remained in this narrative of anthropogenic climate change to allow sceptics numerous opportunities to question cause and effect, but the key point is that developments in the problem stream forced policymakers to engage with the issue at particular points in time. Second, advances in policy knowledge failed to accompany improvements in scientific knowledge (Pielke, 1995; Pielke, 2000a). Even those who accepted that climate change posed a problem that needed to be addressed struggled to offer concrete proposals for action. Scientists attending the Toronto World Conference on the Changing Atmosphere in August 1988 urged governments to establish reduction targets for greenhouse gas emissions for the first time, but on the whole the scientific community preferred to focus on advancing knowledge of the problem rather than possible solutions (Boehmer-Christiansen, 1994a, 1994b). Finally, important changes in the politics stream with long-term consequences occurred during the Reagan years. The emergence of climate change as an international political issue was a significant development as it led to the creation of a process that would periodically force American policymakers to give attention to the issue. The late 1980s also saw environmental interest groups take up the cause of climate change. Climate change became firmly identified as a "green issue" (Weart, 2008, 151).

"No Regrets"

The election of President George H.W. Bush in November 1988 appeared at first sight to promise significant change in US climate change policy. Bush had highlighted the issue during the election campaign, adverse weather events had generated public concern about the prospects of a warming planet, a significant bipartisan consensus had developed in Congress about the need to do something, and international pressures gave added momentum to calls for action. The idea that a "policy window" had opened that would lead to a radical change of policy, however, was misplaced. Divisions existed within the administration about the need for action, no consensus existed in Congress about the best way forward, and major industrial groups had mobilised to resist calls for reductions in greenhouse gas emissions. The early months of the administration did see a change in rhetoric that hinted at a possible embrace of a precautionary approach to the issue with officials suggesting that policy should be guided by the notion of having "no regrets" if climate change turned out to have devastating consequences, but "no regrets" soon came to mean not doing anything rash that might harm the economy or the interests of the energy sector (Pielke, 2000a). Although President Bush signed the United Nations Framework Convention on Climate Change (UNFCCC) in 1992, supported large increases in funding for climate research, and launched a number of other initiatives to improve energy efficiency, the administration's emphasis on economic growth blocked efforts to bring about radical policy change. Proposals to establish binding targets for greenhouse gas emissions, in particular, were rejected by the administration.

The emergence of climate change as an international political issue in the late 1980s meant that President Bush faced a different policy context than his predecessors. Whereas climate policy had previously had an overwhelming domestic dimension the increased involvement of international organisations and governments added a new international dimension to the policy process. Climate policy became a matter of foreign policy as well as domestic policy. This had a number of important consequences for the development of policy. On the one hand, the emergence of an international dimension brought a different set of political actors into the policy process, opened up new venues where policy could be developed, and established a process that required engagement with the issue. Proponents of action hoped that these developments would increase the pressure on the US government to take radical action to address climate change. On the other hand, the emergence of an international dimension added complexity to the policy process, increased the number of political actors involved in the search for solutions, and provided opportunities to avoid taking domestic action. Opponents hoped that the search for an international solution would become mired in endless negotiations between countries and provide an excuse for inaction at the domestic level.

Optimism that President Bush's election would lead to changes in the US government's approach to climate change appeared warranted when Secretary of State James Baker decided to attend a meeting of an IPCC working group in

January 1989. Baker told the meeting that although scientists must continue to "refine the state of our knowledge ... we can probably not afford to wait until all of the uncertainties have been resolved before we act" (Shabecoff, 1989a). He warned that "time will not make the problem go away" and recommended that policymakers should "focus immediately on prudent steps that are already justified on grounds other than climate change," such as improving energy efficiency, planting more trees, and steeper cuts in chlorofluorocarbons (CFCs) than agreed in the Montreal Protocol of 1987. Baker's call for immediate action suggested a break with the policy of the Reagan Administration which had argued that greater research was needed before concrete steps could be taken to address climate change, but behind the scenes American officials continued to demand that the IPCC collect more date about the problem before making any policy recommendations (Cass, 2006, 34).

Disagreements within the Bush Administration about the best way to deal with climate change undermined the type of "no regrets" policy that James Baker had begun to develop in his speech at the IPCC working group (Gold, 1989; Gelb, 1991; Darwall, 2013). While Baker and EPA administrator William K. Reilly favoured action to address climate change, White House chief of staff John Sununu, Office of Management and Budget (OMB) director Richard Darmen, and presidential science advisor D. Allan Bromley advocated a more cautious approach, if not outright scepticism, about the need to do anything (Shanley, 1992; Hopgood, 1998). Two events during spring 1989 illustrate this divide. In March 1989, an EPA draft report to Congress concluded that greenhouse gas emissions were contributing to climate change and made the first specific proposals to address the problem (Shabecoff, 1989b). These included better car fuel efficiency, more fuel efficient homes, planting more trees, imposing fees on fossil fuels to encourage the use of alternative sources of energy, and promoting research on solar power. The EPA warned that these measures would slow but not reverse the build-up of greenhouse gases in the atmosphere. Other elements of the administration took a different view. In May 1989, NASA scientist James Hansen revealed that the OMB had forced him to change his testimony before a Senate committee to make the likelihood of climate change more uncertain (Shabecoff, 1989c). Although the political and public uproar that followed Hansen's accusation forced President Bush to announce that he would convene an international meeting on climate change during autumn 1989, the manoeuvring within the administration over the issue generally favoured the sceptics rather than the proponents of action. One prominent casualty was James Baker, who took the unusual step of recusing himself from further involvement in the development of climate change policy because he had previously worked in the oil industry (Stagliano, 2001, 310; Gore, 2013, 175).

President Bush's early rhetoric on climate change emphasised scientific uncertainty and the need for more research before taking concrete action. In his "Message to Congress Transmitting the Annual Report on International Activities in Science and Technology" on 5 April 1989, for example, he stated that: "Significant uncertainties remain about the magnitude, timing, and regional impacts of global climate change." To reduce these uncertainties Bush initiated the US Global Change Research Program to boost scientific research on global change (Pielke, 2000a; 2000b). In November 1990, Congress formalised the administrative changes made by the president in the Global Change Research Act of 1990. The law established a 10-year plan to "advance scientific understanding of global change and provide usable information on which to base policy decisions relating to global change." International and domestic pressure to do more than conduct research, however, made it difficult for President Bush to maintain his position that more scientific research was needed before action could be taken. The result was a change in rhetoric as Bush sought to frame the issue in slightly different terms. Although his speeches still made frequent reference to scientific uncertainties, Bush also began to talk about the need to evaluate costs and benefits, maintain economic growth, use market mechanisms to achieve goals, and ensure that all countries commit to any international agreement. At the Paris Economic Summit on 14–16 July 1989, President Bush persuaded other leaders to agree to a statement that noted that "Scientific studies have revealed the existence of serious threats to our environment ... which could lead to future climate change" and urged "all countries to give further impetus to scientific research on environmental issues, to develop necessary technologies, and to make clear evaluations of the economic costs and benefits of environmental policies." Six months into his presidency, the guiding principle of "no regrets" had begun to morph from precaution to caution.

American concerns about the economic costs of action to address climate change were not shared (publicly at least) by other developed countries which overwhelmingly framed the issue as an environment problem (Cass, 2006, 41). Part of the explanation for this divergence in approach stemmed from American reliance upon cheap coal to generate power. This meant that the cost of meeting targets for any reductions in greenhouse gas emissions would be higher in the United States than in countries like Canada, France and Sweden, which generated more of their power from nuclear or renewable sources. The political power of industrial groups is another important factor in explaining the American emphasis on economic costs. The fossil fuel industry in the United States had close links with the Bush Administration and senior member of Congress, and used these connections to lobby aggressively against action to address climate change (Layzer, 2007, 99). Industry groups also organised media campaigns to persuade the public that dealing with climate change would be costly. In 1989 the National Association of Manufacturers established the Global Climate Coalition (GCC) to coordinate opposition to growing demands to set targets for greenhouse gas emissions. The GCC not only tried to discredit the scientific basis of climate change, but also funded and publicised a number of studies that purported to show the high costs of reducing greenhouse gas emissions. One consequence was an increase in the number of climate sceptics testifying before congressional committees in the early 1990s (Keller, 2009, 108).

The administration's emphasis on costs and benefits, economic growth, and market mechanisms shaped its domestic response to climate change. When President Bush announced in July 1989 that he had asked the secretary of energy Admiral James Watkins to develop a National Energy Strategy (NES), proponents of action to combat climate change viewed the occasion as an opportunity to reduce America's dependence on fossil fuels (Stagliano, 2001). Two years later, when the NES was finally unveiled, such optimism proved to have been misplaced. Although the NES included some initiatives to promote improved energy efficiency its main thrust was on increasing the supply of fossil fuels (Grossman, 2013, 271). The administration proposed to allow drilling for oil and gas in the Arctic National Wildlife Refuge area, development of five Alaska North Slope oilfields, the lifting of a moratorium on exploration and development of some offshore continental shelf areas, restructuring the electrical generating industry, deregulation of interstate oil and gas pipelines, and the deregulation of the import of natural gas. Substantial elements of the NES were incorporated in the Energy Policy Act of 1992 but Congress added provisions to promote the development of alternative energy sources and improve energy efficiency. Congressional efforts to increase corporate average fuel economy (CAFE) standards for automobiles were rejected by the administration on the grounds of cost. President Bush believed that enactment of the Clean Air Act Amendments of 1990 had already imposed a considerable burden on American industry and was determined to keep the costs of any further environmental action to a minimum (Vig, 1994, 82).

This determination to avoid increasing the regulatory burden on industry led the administration to stress voluntary action to combat climate change. In January 1991 the EPA launched a "Green Lights" programme which provided a range of incentives to persuade businesses to replace old lights with more energy efficient lighting. The EPA claimed that the programme had the potential to reduce carbon dioxide emissions equivalent to removing 44 million automobiles from America's roads. President Bush also proposed to reduce levels of atmospheric carbon dioxide by planting more trees. In his first budget he requested $175 million to plant a billion trees per year. He followed this by submitting legislation to Congress to establish a National Tree Trust that would create public–private partnerships to plant trees. Voluntary action was at the heart of this proposal. In "Remarks at a White House Tree-Planting Ceremony" on 22 March 1990, he claimed that the legislation "will sound a nationwide call for each American to become a volunteer for the environment." The legislation did not pass Congress, but both the Departments of Agriculture and Interior began large scale tree planting programmes (Shanley, 1992, 149).

Concerns about the potential costs of action led to efforts to delay international efforts to negotiate an agreement on climate change despite President Bush's repeated assertions about the willingness of the country to play a leading role in searching for solutions. In a "Statement on International Discussions Concerning Global Climate Change" made on 12 May 1989, Bush claimed that "The United States looks forward to playing a significant role in efforts to assess and respond to global climate change," but such rhetoric was not matched by action. At a UNEP meeting in Nairobi, Kenya, in May 1989, the American delegation rejected calls for work to begin on negotiating a framework convention on climate change, arguing that discussions should wait until the IPCC had reported in 1990 (Cass, 2006, 22). Six months later the American delegation at a UN sponsored conference at Noordwijk, Netherlands, dismissed proposals to stabilise greenhouse gas emissions as soon as possible but did accept a goal of completing a climate change convention in time for the Rio de Janeiro Earth Summit scheduled for 1992. EPA Administrator Reilly had recommended that the administration agree in principle to the goal of stabilising emissions at their existing level sometime in the future, and had secured support from the State Department for this proposal only to be overruled by the White House Policy Council with White House chief of staff Sununu and OMB director Darmen firmly opposed to the idea (Cass, 2006, 36).

President Bush outlined the administration's cautious "no regrets" policy on climate change in February 1990 when he addressed a meeting of the IPCC in Washington, DC. In "Remarks to the IPCC" on 5 February 1990, he asserted that: "The United States is strongly committed to the IPCC process of international cooperation on climate change" and stated that "… the United States will continue its efforts to improve our understanding of climate change." When discussing the possibility of taking action to address climate change he stressed that: "Wherever possible, we believe that market mechanisms should be applied and that our policies must be consistent with economic growth and free-market principles in all countries," and noted that "[o]ur goal continues to be matching policy commitments to emerging scientific knowledge and a reconciling of environmental protection to the continued health of economic development." In these remarks, Bush not only re-stated the long-standing American rejection of precautionary action in the absence of scientific certainty, but also raised newer concerns about economic growth that opened up a need for further research. Proving that proposals for action would not harm the economy required improved knowledge of costs and benefits. To develop this theme further the administration organised a Conference on Science and Economic Research Related to Global Change in April 1990. In "Remarks" to the closing session of the conference on 18 April 1990, President Bush emphasised the need for further economic and scientific research when he claimed: "We're leading the search for response strategies and working through the uncertainty of both the science and economics of climate change." Bush also used the occasion to reiterate his commitment to market mechanisms and rejection of command-and-control regulatory techniques.

A White House memorandum leaked to the press revealed that President Bush's emphasis on scientific and economic "uncertainties" was a deliberate attempt to frame the issue in terms that justified inaction (Shabecoff, 1990). The memorandum counselled that pointing out "the many uncertainties that need to be better understood on this issue" would be a better way of persuading the public of the correctness of the administration's approach than getting involved in a discussion of whether climate change was happening or not. Bush's stress on the need to conduct cost-benefit analysis of the economic impact of potential policy responses formed an important part of this strategy. Identifying and valuing the myriad costs and benefits associated with climate change is extraordinarily problematic given uncertainty over consequences, the timeframe involved, and the difficulty involved in putting a price on environmental features such as ecosystems (Mendelsohn, 2011). The fact that the short-term costs of action are typically easier to specify than the long-term benefits provided opponents of action to address climate change with further ammunition in debates over policy options.

Publication of the IPCC's First Assessment Report in August 1990 prompted renewed international pressure for action to address climate change (IPCC, 1990). Although the report stated that the "unequivocal detection" of anthropogenic climate change would take at least a further decade of observation and warned of "many uncertainties in our predictions," its conclusion that the global mean temperature would rise by 0.3 °C per decade over the next century if greenhouse gas emissions continued unabated led the WMO and UNEP to organise a second World Climate Conference in Geneva in late October 1990 to discuss policy options. European demands for action to stabilise emissions were resisted by the American delegation which argued that further cost-benefit analysis was needed before any new initiatives were undertaken. John Knauss, director of NOAA, told the conference that the United States refused to set greenhouse gas emission targets because we do "not believe in them ... It's as simple as that" (Simons, 1990). The administration's position was that targets were based on political considerations rather than science and few countries had any idea how they would meet them (Darwall, 2013, 139). The Ministerial Declaration issued at the end of the conference largely reflected the American position. It called for more research, stated that action must be taken without delay despite scientific uncertainties, urged developed states to "establish targets and/or national programmes" to limit emissions of greenhouse gases, acknowledged that developing states needed to be able to grow but should still had a responsibility to take action, and called for a framework treaty on climate change to be agreed by the UN Conference on the Environment and Development scheduled for Rio de Janeiro in June 1992 (Condon and Sinha, 2013, 23). The first step towards achieving the later came in December 1990 when the UN General Assembly approved the IPCC's First Assessment Report and created an Intergovernmental Negotiating Committee (INC) to prepare a proposal for a framework convention on climate change (FCCC) in time for the Rio Summit.

The first meeting of the INC took place in Washington, DC, in February 1991. From the outset the administration made clear that it would not accept any agreement that established mandatory greenhouse gas emission targets. Prior to the meeting the White House detailed the American position in "America's Climate Change Strategy: An Action Agenda" (Deland, 1991). This position paper acknowledged that climate change was a problem that needed to be addressed, but argued that the country's *net* greenhouse gas emissions would be stabilised at 1987 levels by initiatives already taken by the United States. These included the removal of CFCs required under the Montreal Protocol of 1987; reductions in stratospheric ozone following from enactment of the Clean Air Act of 1990; proposals for re-forestation included in the 1991 budget; and energy efficiency provisions of the National Energy Strategy. No new policy proposals were included in the position paper. Foreign and domestic proponents of more aggressive action to combat climate change condemned the administration's stance. Delegates to the INC expressed dismay at the American position while Senate Majority Leader George J. Mitchell (D. ME) introduced a resolution in the Senate (S. Res 53) on 7 February 1991 stating that it should be US policy to specify reductions in greenhouse gas emissions by a certain date (Cass, 2006, 79). Such criticism, however, had little effect. The administration remained opposed to the idea of mandatory emission targets.

President Bush outlined the main points of the administration's negotiating position in a "Message to Congress Reporting on Environmental Quality" on 18 April 1991. After stating that "... the United States will continue to seek to conclude an international convention on global climate change in time for its signing at [the Rio Summit]," he proceeded to note what such a treaty should contain. First, Bush argued that the treaty must be "comprehensive in scope." It needed to address all sources and sinks of greenhouse gas emissions, include both adaptation and mitigation responses, and enhance scientific and economic research into the problem. Second, Bush pointed out that the United States was already committed to domestic actions that "will hold United States net emissions of greenhouse gases at or below the 1987 level for the foreseeable future." These included provisions of the Clean Air Act of 1990, energy conservation initiatives, and plans to plant more trees. Finally, Bush concluded that "An effective response to potential climate change also requires that all nations participate and meet obligations that are appropriate to their circumstances." Little was new in Bush's "Message." The basic tenets of the administration's position had been clear for some time, and reflected the views of senior officials that meeting emission targets would restrict economic growth, harm major industries, increase costs to consumers, and require greater government interference in the market. Just a week before President Bush sent his "Message" to Congress his science advisor D. Allan Bromley had dismissed the policy recommendations of an NAS report into climate change on the grounds that "the goals we have in mind are going to be achieved more effectively by people who believe they are doing it for their own benefit or the nation's benefit rather than being forced by some centralised control mechanism" (Weisskopf, 1991).

The administration came under considerable international and domestic pressure to change its position on mandatory greenhouse gas emission reductions in the various meetings of the INC that followed. European countries condemned the administration's approach and both Democrats and Republicans voiced opposition in Congress. In July 1991 Rep. John Porter (R. IL) and Senator Al Gore (D. TN) introduced resolutions (HJ Res 302, SJ Res 181) in the House and Senate that called on the administration to "announce its intention to commit to meaningful reductions in greenhouse gases." Despite this international and domestic pressure, the administration refused to change its position on mandatory emissions targets. On 24 March 1992 President Bush reaffirmed his position in a "Message to Congress on Environmental Goals," which stated that "... an exclusive focus on targets and timetables for carbon dioxide emissions is inadequate to address the complex dynamics of climate change." Rep. Henry Waxman (D. CA) and Senator Gore responded by introducing bills (HR 4750, S 2668) that would require the President "to promulgate regulations to achieve that stabilisation of carbon dioxide emissions by 1 June 2000," but the administration remained resolutely opposed to any mention of mandatory emission targets in a climate change treaty. During negotiations to draft a treaty American officials made clear that President Bush would not attend the Rio Summit if their views were ignored. Faced with such a threat a compromise brokered by the British was eventually agreed that would allow Bush to attend and sign a treaty (Darwall, 2013, 145).

The United Nations Framework Convention on Climate Change recognised that climate change was a problem, and committed industrialised countries to the goal of returning their greenhouse gas emissions to "earlier levels" by the end of the century but contained no legally binding targets. Individual countries were expected to develop their own strategies for addressing the problem depending on their circumstances, and the importance of economic growth, particularly for developing countries, was acknowledged. Other provisions required industrialised countries to support the climate change activities of developing countries, required all countries to provide annual inventories of their greenhouse gas emissions and report on their climate change activities, and established a process to review the implementation of the Framework (an annual "Conference of the Parties"). In a statement made after signing the treaty on 12 June 1992 President Bush praised the Framework Convention as "comprehensive, covering all sources and sinks of greenhouse gases. It provides the flexibility for national programs to be reviewed and updated as new scientific information becomes available." Three months later he urged the Senate to ratify the treaty noting that: "The ultimate objective of the Convention is to stabilise greenhouse gas concentrations (not emissions) in the atmosphere that would prevent dangerous human interference with the climate system." The Senate unanimously approved the Convention on 15 October 1992.

In a brief "Address to the United Nations Conference on the Environment and Development in Rio de Janeiro" on 12 June 1992, President Bush claimed: "We came to Rio with an action plan on climate change. It stressed energy efficiency, cleaner air, reforestation, new technology." Most environmentalists believed that more was needed and criticised the administration's failure to agree to binding cuts in greenhouse gas emissions (Hopgood, 1998). The small steps taken by the Bush Administration to tackle climate change, however, reflected the political forces at play during the early 1990s. A worsening economy and little public engagement with the issue meant a lack of support for radical action to address the problem. The transformation of "no regrets" from a putative acceptance of the precautionary principle to a concern not to harm the economy was in keeping with this political reality. The administration had no enthusiasm for creating new regulatory regimes that would impose additional costs on American industry and preferred to advocate market-led solutions or voluntary action. "… I am confident that the United States will continue to lead the world in taking economically sensible action to reduce the threat of climate change," President Bush noted in his "Statement on Signing the Instrument of Ratification for the United Nations Framework Convention on Climate Change" on 13 October 1992.

Conclusion

The period from the first stirrings of political interest in climate change in the late 1970s to President Bush's signing of the United Nations Framework Convention on Climate Change in 1992 established a pattern of politics that would prove remarkably persistent over the following two decades. First, the debates over climate change revealed themes that would continue to dominate discourse. Arguments about scientific uncertainty, economic harm, bureaucratic regimes, and technological solutions would remain familiar features of debates about whether and what to do about climate change. Second, the early 1990s saw the emergence of a coalition between conservative think tanks and the energy industry willing to challenge claims about the need to address climate change. Manufacturing and disseminating doubt about climate change would become a key strategy of this coalition over the following decades. Finally, the period revealed a lack of public engagement with the issue that would persist despite increases in knowledge. The issue had little saliency and demands for action were easily trumped by other concerns. Where the period differed from the following decades was in the willingness of many Republicans in Congress to join with Democrats to address the issue. Over the next two decades that bipartisan approach would virtually disappear as climate change became one of the wedge issues between the two parties.

Chapter 3

Staggering Towards Kyoto

The election of Bill Clinton in November 1992 offered some encouragement to proponents of more aggressive action to tackle climate change. President Clinton's appointment of several prominent environmentalists to important positions within the administration, and his earlier selection of Senator Al Gore as his running mate, appeared to signal a commitment to dealing with climate change in a more sympathetic way than his predecessor. These hopes of a new approach to the issue proved to be misplaced in the short term. Concerns about the economy, and the early defeat of a proposal for a carbon tax, left the administration offering initiatives that differed little from those of the previous administration. A number of events conspired, however, to shake the administration's approach to climate change out of this default mode. Widespread public hostility to Republican efforts to undermine environmental laws following their success in the midterm congressional elections of 1994 persuaded Clinton that championing the environment might be a vote winner in the 1996 presidential elections; publication of the IPCC's Second Assessment Report in 1996 provided further evidence about the causes and consequences of climate change; and other countries placed considerable pressure on the United States to take action to curb greenhouse gas emissions. In a major departure from existing policy, the administration announced in 1996 that it would accept binding targets for emission targets and approved the Kyoto Protocol in 1997. Further action proved impossible as the administration faced vehement opposition to the Protocol from Congress and powerful industrial groups, and could not persuade other countries to make the compromises necessary to overcome domestic challenges to the idea of emission limits.

The events of the Clinton years show that a "window of opportunity" needed for radical policy change never quite opened. Although President Clinton managed to place the issue on the government's agenda, challenges continued to be aired about scientific certainty, no consensus existed on a solution, and opponents were well placed to challenge the administration's initiatives. Faced with a particular mix of the problem, policy, and politics streams, Clinton simply lacked the power to engineer radical change. Small steps in policy proved possible using his executive authority, but the possibility of legislative action was remote given a lack of public engagement with the issue and widespread congressional concern about the economic costs associated with limits on greenhouse gas emissions.

The 1992 Election

Economic concerns, taxation, health care, and foreign policy dominated the 1992 presidential election with environmental issues in general, and climate change in particular, receiving little attention. The selection of Senator Al Gore (D. TN) as Governor Bill Clinton's running mate in July 1992 thrust the environment into the media spotlight for a few days, but overall neither candidates nor voters regarded climate change as a major issue. The Party Platforms gave little space to the issue, campaign speeches rarely mentioned climate change, and no major exchange of views occurred in any of the three presidential debates. Only in the vice presidential debate on 13 October 1992 was there anything like an extended discussion of the issue. When climate change did emerge during the campaign, however, significant differences were apparent in the positions of the two candidates and parties. The bipartisanship surrounding climate change that had been evident just four years earlier had largely disappeared, and both campaigns sought to frame the issue to their advantage. President Bush's talk of using the "White House effect" to combat the "greenhouse effect" became a distant memory as he derided environmental extremists, caricatured Senator Gore as "Mr Ozone," and emphasised the economic costs of taking action to tackle climate change. The Clinton campaign responded by arguing that millions of jobs would be created by leading "the environmental revolution." This dominant economic frame was in keeping with Bill Clinton's slogan that the election was about "the economy, stupid," but had important consequences long after the 1992 election had been decided. A concern about economic costs would shape policy towards climate change for decades in the future.

The Democratic National Convention in July 1992 provided the first concrete sign that a divide on climate change might become apparent during the 1992 presidential election. First, Governor Clinton announced that Senator Gore would be the party's candidate for vice president. Gore had a long record of championing action to address climate change in Congress, had been vocal in criticising the Bush Administration's approach to the Rio Summit, and had just published a book *Earth in the Balance*, which discussed the need to tackle global environmental problems (Gore, 1992). The contrast with Vice President Dan Quayle was sharp and noted by the media (Rosenbaum, 1992). Vice President Quayle had chaired the White House Council on Competitiveness and in that capacity had proposed numerous changes to environmental regulations to reduce the burden on American businesses. Second, the Democratic Party Platform published on 13 July 1992 stated that "We should join our European allies in agreeing to limit carbon dioxide emissions to 1990 levels by the year 2000." This commitment to a binding target for reducing greenhouse gas emissions stood in stark contrast to the Bush Administration's position. The Republican National Convention in August 1992 confirmed the gap between the Democrats and Republicans on climate change. In his "Address to the Convention" on 17 August 1992, Patrick Buchanan told the delegates that at the Democratic National Convention "Mr Gore made a startling declaration. Henceforth, Albert Gore said, the 'central organizing principle' of government everywhere must be: the environment. Wrong, Albert. The central organizing principle of this republic is: freedom ... America's great middle class has got to start standing up to these environmental extremists who put birds and rats and insects ahead of families, workers and jobs." The Republican Party Platform published the same day made a similar point. "[W]e applaud our President for personally confronting the international bureaucrats at the Rio Conference," the Platform stated, "He refused to accept their anti-American demands for income redistribution, and won instead a global climate treaty that relies on real action plans rather than arbitrary targets hostile to US growth and workers." The Platform also promised that: "... a Republican Senate will not ratify any treaty that moves environmental decisions beyond our democratic process or transfers beyond our shores authority over US property."

Disputes about economic costs and jobs losses dominated the few exchanges on climate change during the election campaign. In the presidential debate on 19 October 1992, President Bush claimed that an earlier commitment by Bill Clinton to increase Corporate Average Fuel Economy (CAFE) standards "would break the auto industry and throw a lot of people out of work." Clinton responded that he had made no such commitment and had merely indicated a goal of raising the standard to 45 mpg. Bush returned to the attack in late October 1992. In a "Question-and-Answer Session" in Grand Rapids, Michigan, on 29 October 1992 he defended his administration's position on climate change as necessary to protect American jobs. "On climate change, we did change it a little bit, because I don't want to see us burden the automotive industry with the kind of costs that the Europeans wanted us to put on the industry," he stated in response to a question about his environmental record, "... we can't go off to the extremes and still talk about how we're going to help all those people that are looking for jobs." Labelling environmentalists as extremists formed a key element of Bush's rhetorical strategy much as he had caricatured his Democratic opponent Governor Michael Dukakis as a "liberal" in the 1988 election campaign. Throughout the 1992 campaign, President Bush referred to Senator Gore as "Mr Ozone" in an effort to ridicule his concern about climate change. Defending his attacks on Gore in "Remarks to the Community" in Sussex, Wisconsin, on 31 October 1992, he stated: "They get on me about calling Senator Gore Mr Ozone. Well, let me tell you what I mean ... We've got a good record. But jobs matter. Families matter. Jobs and families ought to take a little priority around here, if you ask me."

Differences between the two campaigns on climate change were highlighted in the vice presidential debate on 13 October 1992 when Vice President Quayle and Admiral James Stockdale (independent candidate Ross Perot's running mate) attacked Senator Gore on the issue. Arguments about job losses, taxes, and costs dominated the early exchanges. Quayle claimed that a Clinton Administration would increase CAFE standards and introduce energy taxes with disastrous consequences for American workers. "You ought to ask somebody in Michigan, a UAW worker in Michigan, if they think that increasing the CAFE standards ...

to 45 miles a gallon is a good idea—300,000 people out of work," he stated, "You ought to talk to the coal miners. They're talking about putting a coal tax on. They're talking about a tax on utilities, a tax on gasoline and home heating oil—all sorts of taxes." Admiral Stockdale also raised a question about costs when he commented "I read Senator Gore's book about the environment and I don't see how he could possibly pay for his proposals in today's economic climate." Gore attempted to deflect such concerns by dismissing claims that environmental protection leads to job losses. "Bill Clinton and I believe we can create millions of new jobs by leading the environmental revolution instead of dragging our feet and bringing up the rear," he stated, "We cannot stick our heads in the sand and pretend that we don't face a global environmental crisis, nor should we assume that it's going to cost jobs." Later exchanges sought to question the scientific basis of climate change. Referring to a paper that Roger Revelle had co-authored with Fred Singer that questioned the scientific certainty of climate change (Singer et al., 1991), Stockdale asked Gore: "I read where Senator Gore's mentor had disagreed with some of the scientific data that is in the book [*Earth in the Balance*]. How do you respond to these criticism of that sort?" Gore explained that Revelle had been very ill when the paper had been drafted and probably didn't fully understand what had been written (Oreskes and Conway, 2010, 190–96).

No evidence exists that the debates over climate change influenced the outcome of the 1992 election. Bill Clinton won the election because he persuaded voters that he had better ideas about how to respond to the economic recession than President Bush. This does not mean, however, that the debates about climate change were unimportant. First, the rhetoric and issues raised during the campaign established the frames that would dominate policy debates for the next two decades. Questions about science and concerns about the economic costs of action constituted the most important of these frames, but the 1992 election campaign also saw efforts to portray climate change in terms of big government and threats to American independence. Second, the 1992 election provided evidence of a partisan split over climate change that would become steadily wider over the subsequent two decades. Republican elites and supporters increasingly took sceptical positions on climate change while their Democratic counterparts moved in the opposite direction.

Further Small Steps

Proponents of aggressive action to address climate change welcomed Bill Clinton's victory in the 1992 presidential election (McCarthy, 1992; Dowie, 1995, 177). Although Clinton had placed little emphasis on environmental issues as Governor of Arkansas, his selection of Senator Al Gore as his running mate, and the appointment of a number of people with impeccable environmental credentials to his administration, suggested a new approach to climate change might be in the offing. Key appointments included Carol Browner, head of

Florida's Department of Environmental Regulation and a former aide to Senator Gore, as EPA administrator; Bruce Babbitt, former governor of Arizona and president of the League of Conservation Voters as secretary of the interior; Senator Tim Wirth (D. CO) as under secretary of state for global affairs; and Rafe Pomerance of the World Resources Institute to lead the American negotiating team at the INC. Not all members of the administration, however, regarded the environment or climate change as a priority. Treasury Secretary Lloyd Bentsen and Energy Secretary Hazel O'Leary, in particular, argued against setting targets for reducing greenhouse gas emissions. They argued that insufficient studies had been conducted on the effect that such reductions would have on the economy (Schneider, 1993). Divisions within the administration on climate change were evident from the start and undermined the prospects for radical policy change in the short term. Key political actors expressed scepticism about the need to take action and no consensus existed on the appropriate solution. Within months of taking office the new administration sounded and acted much like its predecessor.

Early indications suggested that the Clinton Administration would promote a different approach to climate change than that pursued by the Bush Administration. In his "Address Before a Joint Session of Congress on Administration Goals" on 17 February 1992, President Clinton failed to mention climate change, but did propose a broad-based tax on energy as a means of raising revenue to reduce the budget deficit and promote energy efficiency. Environmentalists viewed the proposal as a first step to reduce greenhouse gas emissions using taxes (Lippman, 1993). Republican opposition, unease among many Democrats, and intense lobbying by energy producers and manufacturers, however, ensured the defeat of the measure in the Senate. Instead of a broad-based tax on energy the Senate forced Clinton accept a modest increase in the tax on gasoline to generate revenues (Cass, 2006, 98).

Two months after his address to Congress, President Clinton signalled an apparent departure from the policy of his predecessor when he announced in his "Remarks on Earth Day" on 21 April 1993 that: "I reaffirm my personal, and our Nation's commitment to reducing our emissions of greenhouse gases to their 1990 levels by the year 2000." Clinton stated that he had instructed officials to produce a "cost-effective plan by August that can continue the trend of reduced emissions." A close reading of President Clinton's "Remarks on Earth Day," however, reveals that enthusiasm for radical policy change had already disappeared. In language similar to that used by President Bush, Clinton argued that: "[O]nly a prosperous society can have the confidence and means to protect its environment," and that, "our policies must protect our environment, promote economic growth, and provide millions of new high-skill, high-wage jobs." Referring specifically to climate change, Clinton's instructions to officials to produce an action plan by August 1993 warned that: "This must be a clarion call, not for more bureaucracy or regulation or unnecessary costs but, instead, for American ingenuity and creativity, to produce the best and most energy-efficient technology." The references to economic growth, cost effectiveness, limited government, and technological innovation in

Clinton's "Remarks" provide evidence of the success that Republicans had in framing debate on climate change during the Bush Administration. Just four months into his administration, Clinton was using essentially the same tropes as his predecessor to discuss and shape policy options. The only concrete policy changes announced in the "Remarks" were executive orders to increase the number of clean-fuel automobiles (EO 12844) and energy-efficient computers (EO 12845) purchased by the federal government. Further small steps in policy rather than large strides appeared to be the order of the day.

The eventual release of the administration's climate action plan in October 1993 confirmed that plans for radical changes in policy had been abandoned and replaced by proposals little different than those pursued by the Bush Administration. The action plan emphasised public–private initiatives, called for companies and individuals to take voluntary action to improve energy efficiency, but contained no new regulatory measures or taxation to reduce greenhouse gas emissions. In "Remarks at the White House Conference on Climate Change" on 19 October 1993, President Clinton introduced the action plan in terms that could have been used by his predecessor. Clinton claimed that the plan "seeks to give the American people the ability to compete and win in the global economy while meeting our most deep and profound environmental challenge" by "harnessing private market forces … to bring out our best research and our best technologies." Voluntary action constituted a key component of the action plan. "So I say to the American people: if your utility offers you help in conserving energy in your home, seize it. If you own a business and the EPA offers you a chance to join the Green Lights program, do it," Clinton counselled in his "Remarks." Environmentalists criticised the emphasis of voluntary action and complained about the lack of any proposals for mandatory emissions reductions (Lee, 1993).

Fear of being accused of harming the economy, adding to the budget deficit, and fostering big government shaped the development of climate change policy during the initial years of the Clinton Administration. In "Remarks on the Observance of Earth Day" on 21 April 1994, President Clinton took care, for example, to note that the administration's actions to protect the environment had not created bureaucracies "that grow faster than weeds" and had reinvented "the way we protect the environment so that government is a partner, not an overseer." Defeat of proposals to introduce a broad-based energy tax in the early months of the administration added a further wariness of ambitious proposals that required congressional approval. Using existing statutory authority to make incremental changes in policy proved the preferred option. In a congressional committee hearing in May 1994, Energy Secretary Hazel O'Leary told senators that the Energy Policy Act of 1992 was "the keystone of [the administration's] Climate Action Plan" (SCENR, 1994). The focus on voluntary action to improve energy efficiency contained in the action plan followed from the decision to use the 1992 law as the basis for the administration's actions. "Our climate change programs help companies and consumers save energy and money with air conditioners, computers, and light bulbs that use less electricity than ever before. And we're helping American companies to build those products and create those jobs," President Clinton stated in his Earth Day "Remarks" in 1994.

Republican victories in the midterm elections of November 1994, which gave the party majorities in both the House of Representatives and the Senate for the first time since 1954, effectively ended immediate prospects of a major change in domestic climate change policy. The new Republican majority, particularly in the House under the leadership of Speaker Newt Gingrich (R. GA), regarded environmental protection as a prime example of the big government they had campaigned to cut during the election (Drew, 1997). Majority Whip Tom Delay (R. TX), for example, stated in a 1995 speech on the floor of the House of Representatives that: "The EPA, the gestapo of government, pure and simply has been one of the major claw-hooks that the government maintains on the back of our constituents" (Gerstenzang, 1995). Government action to combat climate change proved a particular target of the new majority's zeal to roll-back environmental protection. Republicans took advantage of their institutional power to organise congressional committee hearings to give opportunities for climate change sceptics to voice their concerns about science, and sought to reduce or cut-off funding for further research and action using the budget and appropriations process. This assault forced the Clinton Administration onto the defensive and left little space for new domestic initiatives. President Clinton was forced to use his veto to defend programmes from budget cuts and had to rely upon existing regulatory authority to take any new action. Over the next six years Clinton issued a number of executive orders to require federal facilities to conduct energy audits (EO 12902), purchase more alternative fuelled vehicles (EO 13031), require federal agencies to reduce their greenhouse gas emissions by 30 per cent by 2010 (EO 13123), and reduce the federal government's annual consumption of petrol by 20 per cent by 2005 (EO 13149).

Republicans signalled their intention to question the science of climate change early in the 104th Congress (1995–1996) when Rep. Dana Rohrabacher (D. CA), chair of the Committee on Science's Subcommittee on Energy and the Environment, announced that the panel would hold hearings to investigate allegations of political interference in the science underpinning policy on dioxins, ozone depletion, and climate change (McCright and Dunlap, 2003; Oreskes and Conway, 2010). The hearing on climate change, held in November 1995 to pre-empt the findings of the IPCC's 2nd Assessment Report, gave an opportunity for climate sceptics to air grievances about the exclusion of scientists who questioned the dominant view of anthropogenic climate change from international scientific assessment bodies, and a chance to question the validity of the scientific consensus on the condition (HCS, 1995). Subsequent committee hearings in both the House of Representatives and the Senate over the next six years continued in this vein. Republicans gave considerable opportunities for climate sceptics to challenge the scientific consensus on climate change, dismiss evidence that warming had been detected, posit that natural causes accounted for any warming that had been detected, suggest that no adverse consequences would result from climate change, and argue that continued scientific uncertainty precluded policy action (Keller, 2009). The hearings provided a clear statement of the dominant Republican view of climate change during the last decade of the twentieth century, and served to legitimise their efforts to cut-back federal programmes designed to address the problem.

Legislation to reduce or eliminate funding for federal climate change programmes was introduced by Rep. Rohrabacher and Rep. Robert Walker (R. PA) in 1995. Both the Environmental Research, Development and Demonstration Authorization Act of 1995 (HR 1814) introduced by Rohrabacher and the Omnibus Civilian Science Authorization Act of 1995 (HR 2405) included provisions to cut-off funding for the Climate Action Plan. The latter passed the House of Representatives but no action was taken in the Senate. Provisions stripping funding were also included in budget legislation and appropriation bills passed by Congress prompting vetoes from President Clinton. Clinton vetoed the Department of Veterans Affairs, Housing and Urban Development, and Independent Agencies Appropriations Act for Fiscal Year 1996 in December 1995, for example, because it cut funding for climate change programmes among other things. Similar efforts to eviscerate research into climate change and reduce government support for the public–private initiatives contained in the administration's Climate Action Plan continued in the 105th Congress (1997–1998) and the 106th Congress (1999–2000), forcing President Clinton to veto, or threaten to veto, major spending bills to protect programmes. Republicans even made an attempt to restrict the EPA's ability to communicate information about climate change in the Department of Veterans Affairs, Housing and Urban Development, and Independent Agencies Appropriations Act for Fiscal Year 1999, but the measure was eventually rejected by the House. President Clinton commented on the attempt in a "Statement on House of Representatives Action on Environmental Legislation" made on 23 July 1998 when he stated: "… the American people expect and deserve a fair, honest, and informed debate on the issue of climate change. Some in Congress would have stifled that debate by effectively imposing a gag order on federal agencies. Thankfully, the House voted to remove this language from the VA-HUD appropriation bill."

Efforts to address climate change did not completely disappear in Congress during the late 1990s despite the opposition of the conservative Republican majority. Senator John Chafee (R. RI) had an interest in the problem that dated back to the 1980s and used his position as chair of the Senate's Environment and Public Works Committee to introduce legislation in the 105th and 106th Congresses to provide incentives for businesses to reduce emissions of greenhouse gases. Both the Credit for Voluntary Action Act (S. 2617) introduced on 10 October 1998 and the Credit for Voluntary Reduction Act (S. 574) introduced on 4 March 1999 proposed to amend the Clean Air Act of 1990 to provide regulatory credit for early action to mitigate greenhouse gas emissions. Other proposals sought to use the tax system to provide incentives for voluntary action to reduce emissions. Senator Larry Craig (R. ID) introduced the Climate Change Tax Amendments of

1999 (S. 1777) in the 106th Congress for this purpose. Legislation introduced by Senator Sam Brownback (R.KS) in the 106th Congress also proposed providing tax incentives to boost voluntary action to store carbon (S. 2540; S. 2982). In contrast to this Republican emphasis on voluntary action, Democratic initiatives proposed regulatory action to establish emission limits. The Clean Power Plant and Modernization Act of 1999 (S. 2626), introduced by Senator Patrick Leahy (D. VT) in the 105th Congress (and re-introduced in the 106th Congress as S. 1949), proposed to establish emission limits for carbon dioxide based on efficiency standards. Rep. Tom Allen (D. ME) also introduced legislation in the 106th Congress to limit emissions of carbon dioxide from power stations but offered a different means of doing so. The Clean Power Plant Act of 1999 (HR 2980) proposed to give allowances to power plants to emit specific amounts of carbon dioxide and required them to return unused allowances to the government.

No legislative action took place on any of the bills introduced in the 105th and 106th Congresses to tackle climate change as the majority remained resolutely opposed to their consideration. The fact that the measures were introduced in such a hostile legislative environment, however, shows that the problem still attracted the attention of a number of legislative entrepreneurs. Some of these had long been interested in climate change. Senator Chafee had worked with Senator Gore and Senator Wirth in the 1980s to fashion early responses to the issue, but would die in October 1999 shortly after announcing that he would not seek re-election in 2000. Others showed interest for the first time. In this context, the emerging interest of Senator John McCain (R. AZ) in climate change was significant. Late in the 106th Congress, McCain introduced the International Climate Change Science Commission Act (S. 3237) which proposed providing support and resources for an international commission to assess climate change and provide policy options. Nothing came of the proposal but it signalled that an important member of the Republican establishment had different views of the way forward than the party's presidential candidate Governor George W. Bush. Following Bush's election in November 2000 these differences would become stark as McCain became a leading advocate of cap-and-trade as a means to reduce greenhouse gas emissions.

Moving Towards Kyoto

The emergence of climate change as an international political issue during the late 1980s and early 1990s generated momentum for policy action independent of the domestic political concerns facing the Clinton Administration. First, the IPCC process continued to gather and assess scientific information about the problem and its potential consequences (Bolin, 2007). IPCC reports served to remind policymakers that the majority of scientists believed a problem existed and to force engagement with the issues raised (even if only to dismiss the IPCC findings). Second, the United Nations Framework Convention on Climate Change (UNFCCC) agreed at the Rio Summit in June 1992 created an institutional process to review the action taken by countries to address climate change and negotiate revisions to the treaty if needed. The Intergovernmental Negotiating Committee (INC) that had crafted the UNFCCC continued to meet until the Convention came into force in March 1994 when preparations began for the first Conference of the Parties (COP-1) scheduled to be held in Berlin a year later. This process required continued American involvement in international deliberations about climate change even if the issue lacked saliency in the domestic arena.

Early actions of the Clinton Administration appeared to promise a greater commitment to working with other countries to address climate change than had proved the case during the Bush Administration. The appointments of former Senator Wirth and Rafe Pomenrance to positions in the State Department, with responsibility for negotiating with the INC, and President Clinton's 1993 Earth Day Remarks in which he stated that the administration would adopt a programme to implement the emission reduction targets contained in the UNFCCC and promised that we "must take the lead in addressing the challenge of global warming," suggested a departure from the obstructive attitude of the previous administration. The climate action plan announced by President Clinton in October 1993, however, revealed a policy approach that differed little from that of the Bush Administration. The plan contained no proposals for mandatory emissions reductions, relied heavily on voluntary action, and called for joint implementation (where one country can obtain "credit" to continue polluting by paying for emission reduction activities in another country) as a cost-effective way of reducing global greenhouse gas emissions that embraced market mechanisms. Opposition to emission reduction targets and timetables, demands that developing countries engage meaningfully in emission reduction activities, and an insistence on joint implementation, dominated the American negotiating position in the post-Rio INC negotiations much like they had in previous years.

Few industrial countries believed that the commitments to mitigate greenhouse gas emissions that had been made since approval of the UNFCCC in June 1992 were adequate to meet the treaty's objectives, and agreed that a new Protocol was needed to remedy short-comings. Even the United States concurred with this view (Cass, 2007, 117; Royden, 2002, 412–22). The problem was a lack of consensus about what to do. United States negotiators at an INC meeting in August 1994 floated the idea of a "new aim" that would cover the post-2000 period but failed to specify what it should be, suggested that "advanced" developing countries should make more of an effort to reduce their greenhouse gas emissions, and vigorously promoted the idea of joint implementation (Cass, 2007, 118–19). The European Union urged the adoption of a new protocol that would cover all greenhouse gases, while China argued that no new commitments should be agreed until existing ones had been met. Eventually, the Germans suggested that COP-1, scheduled to take place in Berlin in March 1995, should be used to negotiate a mandate for a new protocol to be agreed in time for COP-3 in 1997. The INC agreed with this proposal to postpone discussions.

The Clinton Administration reiterated well-worn arguments in the negotiations leading to COP-1. In language reminiscent of that used by the Bush Administration prior to the Rio Summit in 1992, officials opposed setting targets and timetables, demanded greater commitment from developing countries, and called for further analysis of the economic costs and benefits of the various policy options available (Cass, 2007, 120). Rafe Pomerance told a congressional committee just before COP-1 that the administration supported a "structure" but not specific targets and timetables, and a "mandate" for negotiations rather than a regulatory mandate (Hopgood, 1998, 215–16). The negotiating positions of other countries was equally familiar with Europeans generally calling for legally binding emission targets, and developing countries such as China and India opposing demands for action that would slow their economic growth. Intense negotiations in Berlin eventually produced a compromise agreement that established important parameters for a new Protocol to be agreed at COP-3 scheduled to take place in Kyoto in December 1997. The Berlin Mandate stated that more action needed to be taken by developed countries to meet the objectives set out in the UNFCCC, but that no new commitments were required from developing countries to mitigate greenhouse gas emissions. An Ad Hoc Group was established to negotiate emission reductions within specified time frames albeit without a reference to any target base line. To secure American support the Berlin Mandate also established a pilot project for Joint Implementation.

The American delegation at COP-1 dismissed European claims that the Berlin Mandate required reductions in greenhouse gas emissions from 2000 onwards. US Under Secretary of State Wirth argued that we "agreed there are to be reductions, but the word 'target' does not appear and there is no specific timetable ... I don't think there's a baseline in this" (Liffey, 1995). Wirth repeated his view that the Berlin Mandate contained no commitments beyond agreeing to participate in negotiations in a congressional committee hearing in May 1995 (HCC, 1995). Few members of the congressional committee concurred in this interpretation of what had been agreed in Berlin. Both Republicans and Democrats feared that the administration had committed the United States to a process that would require expensive emission reductions which would place the country at an economic disadvantage because developing countries such as China would have no such commitments. American industry soon added its voice to the debate. The Global Climate Coalition and other industrial groups engaged in an intense lobbying and public relations campaign following COP-1 to highlight the costs and loss of competitiveness that they believed would follow any agreement to reduce greenhouse gas emissions (Lewis, 1995).

Although the Clinton Administration continued to maintain that the Berlin Mandate had established a process rather than a regulatory outcome, two developments conspired to produce a shift in its position on greenhouse gas emissions targets over the following months. First, evidence began to emerge that Republican efforts to weaken environmental laws following the 1994 midterm elections had re-ignited public concern about the environment. This presented

President Clinton with an opportunity to court voters in the run-up to the 1996 presidential election by stressing his environmental credentials and attacking Republicans for their efforts to sabotage laws. In "Remarks on Environmental Protection" made on 8 August 1995, Clinton expressed surprise at "this dramatic departure from the bipartisan efforts of the last 25 years" to protect the environment, and claimed that Republicans had caved in to the "pressure of lobbyists with vested financial interests in seeing that happen." Second, the IPCC released a draft of its Second Assessment Report (SAR) in December 1995 which concluded that "the balance of evidence suggests that there is a discernible human influence on global climate" (IPCC, 1996). This finding made it more difficult to argue that the science did not support action to address climate change. While opinion polls suggested that President Clinton might benefit politically by displaying a stronger commitment to environmental protection, SAR provided a justification for a shift in a specific environmental policy. The administration calculated that agreeing to legally binding targets for greenhouse gas emissions would not harm it politically if certain conditions were met (Royden, 2002; Hopgood, 1998).

Under Secretary of State Wirth announced the administration's willingness to negotiate a binding protocol containing targets and timetables at COP-2 in Geneva in July 1996 (Downie, 2014; Meckling, 2011). Wirth's statement began by expressing acceptance of the IPCC's findings. Dismissing efforts to undermine SAR, he declared that: "The science calls upon us to take urgent action; the IPCC report is the best science we have, and we should use it." Wirth proceeded to outline three principles that needed to underpin this urgent action. He stated that outcomes must be "real and achievable" rather than overly ambitious, that use should be made of "market-based solutions that are flexible and cost-effective" and "all countries—developed and developing—must contribute to the solution to this challenge." Little was new about these principles. What was new was the commitment that followed. Wirth told delegates that: "Based on these principles ... the United States recommends that future negotiations focus on an agreement that sets a realistic, verifiable and binding medium-term emissions target." This was the first public statement that the Clinton Administration would support legally binding targets for greenhouse gas emissions though no information was given about what they might be. In an interview with a reporter from *The New York Times*, Wirth underscored the importance of his statement: "This is a big deal ... Saying that we want to have a target that is binding is a clear indication that the United States is very serious about taking steps and leading the rest of the world" (Cushman, 1996).

Environmental groups welcomed the administration's decision to support legally binding emissions targets but industrial groups reacted swiftly to warn of the economic costs that would follow (Lee, 1996). Republicans also condemned the proposal. The Republican Party Platform published on 17 August 1996 contained a lengthy rejection of the Clinton Administration's policy that raised familiar concerns about scientific uncertainty, economic costs, and loss of sovereignty. The Platform stated that: "Despite scientific uncertainty about the role of human activity in climate change, the Clinton Administration has leapfrogged over reasoned scientific inquiry and now favours misdirected measures, such as binding targets and timetables, imposed only on the United States and certain other developed countries, to further reduce greenhouse gas emissions. Republicans deplore the arbitrary and premature abandonment of the previous policy of voluntary reductions of greenhouse gas emissions. We further deplore ceding US sovereignty on environmental issues to international bureaucrats and other foreign countries." President Clinton lessened the impact of such attacks by keeping things vague. The Democratic Party Platform published on 26 August 1996 simply stated that: "We will seek a strong international agreement to further reduce greenhouse gas emissions worldwide and protect our global climate." This refusal to provide details of targets or timetables undermined the ability of Republicans and industry groups to mobilise opinion against the administration's climate change policy as there were no specifics to challenge (Cass, 2006, 126).

Climate change failed to materialise as a prominent issue in the 1996 presidential election. Not President Clinton, Senator Bob Dole, or Ross Perot gave a major speech on the issue, and exit polls suggest that voters were concerned about crime, the budget deficit, and the economy (Roper, 1996). President Clinton won the election with 49 per cent of the popular vote, and Democrats made gains in Congress, though the Republicans maintained their majority. The results failed to alter the political dynamics of climate change. Key players involved in climate change negotiations, such as Timothy Wirth and Rafe Pomerance, remained in important positions; Republicans continued to enjoy a majority in Congress; industrial groups and conservative think tanks had the resources to undertake lobbying and public relations campaigns; and the general public had other priorities. Despite the IPCC's findings, enough doubt about the evidence of anthropogenic climate change remained to allow sceptics opportunities to challenge the science, and no consensus existed about how to respond to the problem. Congressional Republicans and Democrats, in particular, remained deeply concerned that mandatory emission limits would harm the American economy. Rep. John Dingell (D. MI), the ranking member of the Commerce Committee, voiced the fears of many when he stated before the election that: "If this is handled badly it could help promote the deindustrialization of the United States" (Morgan, 1996).

President Clinton sought to counter opposition to mandatory emission limits by emphasising the costs of not taking action. In "Remarks During a Discussion on Climate Change" on 24 July 1997, he noted that: "… the overwhelming balance of evidence and scientific opinion is that it is no longer a theory but a fact that global warming is for real," and warned that a failure to act would lead to a rise in sea levels that would leave large areas of Florida and Louisiana underwater, an increase in infectious diseases, severe heat waves, more frequent floods and droughts, and a decline in agricultural output. Clinton acknowledged that there was a large gap between scientific claims and the public's experience of climate change, but sought to persuade Americans that they had a moral duty to protect future generations from the consequences of climate change. He declared that "[w]e have the evidence, we can see the train coming, but most ordinary Americans in their day-to-day lives can't hear the whistle blowing," and conceded that "... the degree of the challenge is inconsistent with the actual perceived experience of most ordinary Americans," but argued that "we have to see this whole issue of climate change in terms of our deepest obligations to future generations." This obligation, however, was not unconditional. Clinton finished by reiterating the administration's terms for reaching agreement at Kyoto. He stated that any binding limits on greenhouse gas emissions had to be realistic, that market-based approaches and technological innovation needed to be central to the way that such targets were met as this "will help us improve the economy," and that "we have to ask all nations ... to participate in this process."

The size of the task confronting President Clinton in his efforts to sell binding emissions limits became apparent a day later when the Senate passed the Byrd-Hagel Resolution (S. Res 98) by a vote of 95–0. This Resolution expressed the sense of the Senate that the United States should not sign any protocol at Kyoto that mandated greenhouse gas emission reductions unless developing countries also made new commitments to reduce their emissions, or would result in "serious harm to the economy." A similar resolution was introduced in the House of Representatives by Rep. Joe Knollenberg (R. MI), but not acted upon. Although Byrd-Hagel had little legal consequence as the US Constitution clearly gives the president the power to negotiate treaties, the fact that the Senate would have to ratify any agreement signed at Kyoto sent a powerful political message to the administration about what sort of agreement would be acceptable to senators. Senator Robert Byrd (D. WV) explained that: "I do not think the Senate should support a treaty that requires only ... developed countries to endure the economic costs of reducing emissions while developing countries are free to pollute the atmosphere, and, in so doing, siphon off American industries" (Dewar, 1997). Concern for the jobs of West Virginia's coal miners also motivated Byrd.

The administration's climate change policy received some support from large companies such as DuPont, General Electric, and AT&T, and some members of the insurance industry who were concerned about the prospect of catastrophic loses if predictions of the consequences of climate change proved true, and the alternative energy industry, but faced determined opposition from most of the business community (Layzer, 2007, 103; Kolk and Levy, 2001). Groups representing the automobile manufacturers, the oil industry, and farmers launched a major public relations campaign which warned of the economic problems that would flow from efforts to curb greenhouse gas emissions (Cass, 2006, 129). The Global Climate Coalition (an umbrella group of 54 industry groups) spent $13 million on a television advertising campaign that predicted rising energy prices, increases in the cost of living, and job losses in the car and mining industries if strict reductions in greenhouse gas emissions were agreed (Christiansen, 1999, 258). Other advertisements asserted that developing countries would not have to make cuts in their greenhouse gas emissions and told viewers that "... while the United States is forced to make drastic cuts in energy use, countries like India, China, and Mexico are not" (Kurtz, 1997). Conservative think tanks such as the Competitive Enterprise Institute reinforced this message with a range of activities that sought to discredit the science of climate change (McCright and Dunlap, 2000). Materials promoting sceptical views of climate change were produced and circulated through press releases, press conferences, sponsored events, and radio and television interviews.

President Clinton made an attempt to counter, or assuage, the arguments of opponents in "Remarks at the National Geographic Society" on 22 October 1997 when he announced the administration's negotiating position for Kyoto. He began by stressing that "the problem is real. And if we do not change our course now, the consequences sooner or later will be destructive for America and for the world." Climate change, he warned, would lead to an increasing number of disruptive weather events, the movement of disease-bearing insects into new areas, rising temperatures, and the retreat of glaciers. Dismissing the arguments of industry, Clinton claimed that meeting the challenge of climate change would "create a wealth of new opportunities for entrepreneurs at home" and create new jobs particularly in the energy efficiency and clean energy sectors. "If we do it right, protecting the climate will yield not costs but profits, not burdens but benefits, not sacrifice but a higher standard of living," he stated. Clinton provided an indication of what was involved in "doing it right" by outlining proposals for Kyoto and offering a package of domestic initiatives to tackle climate change. He stated that any new international agreement should require industrialised countries to accept "the binding and realistic target of returning to emissions of 1990 levels between 2008 and 2012," "embrace flexible mechanisms for meeting these limits," such as emissions trading and joint implementation, and warned that the United States "will not assume binding obligations unless key developing countries meaningfully participate in this effort." Clinton also announced a number of domestic initiatives to tackle climate change that included proposals for tax cuts and government spending to promote investment in clean energy technologies, creating incentives for businesses to take voluntary action, bringing competition to the electricity industry, and action to reduce the carbon footprint of the federal government.

The proposals outlined by President Clinton reflected the international and domestic political pressures bearing down on the administration at the time. On the one hand, European leaders wished to secure an agreement at Kyoto that committed industrialised countries to significant reductions in greenhouse gas emissions. In May 1997 the European Union proposed that the new treaty should require all industrialised countries to reduce their greenhouse gas emissions by 15 per cent of 1990 levels by 2010. Clinton came under considerable personal pressure at the G-8 Summit in Denver in June 1997 to agree to European demands and later told an oral historian that he had been "upbraided" on the issue by the country's allies (Darwall, 2013, 172; Royden, 2002, 434). On the other hand, domestic political circumstances acted as a barrier to aggressive action on greenhouse gas emissions. The Byrd-Hagel Resolution provided evidence of significant opposition in Congress to any treaty perceived to damage America's economic interests, business and conservative groups had the resources to launch a determined lobbying campaign to defeat an agreement they did not like, and the public lacked engagement with the issue. The proposals announced by President Clinton in October 1997 sought to "walk a fine line" between these competing pressures (Cass, 2006, 133). They represented an effort to juggle European demands for an ambitious treaty that would be rejected domestically and calls for limited action that would alienate some of the country's key international partners. Robert Putnam (1988) has described such negotiations as a "two-level game."

The negotiations in the months leading to COP-3 in Kyoto in December 1997 proved that "walking a fine line" or playing the "two-level game" is extremely difficult. Europeans refused to compromise on emission targets and developing countries such as China remained opposed to making any commitments that might damage their economic growth. Little changed when negotiations began in Kyoto. Faced with the collapse of negotiations and the prospect of being blamed for their failure, Vice President Gore flew to Kyoto and issued instructions to the American negotiating teams "to show increased negotiating flexibility if a comprehensive plan can be put in place" (Stevens, 1997). Gore's intervention provided the impetus to reach an agreement. The United States agreed to reduce greenhouse gas emissions by 7 per cent below 1990 levels between 2008–2012 in return for a number of compromises on flexibility and joint implementation. No compromise proved possible, however, on the question of the participation of developing countries in mitigation efforts. "We have said categorically *no*" was the position of G-77 countries (Barnum, 1997). Although the agreement received generally favourable press coverage in the United States, it was condemned by key industrial groups and garnered only lukewarm support from a number of environmental groups (Royden, 2002).

Stumbling After Kyoto

Debate over the direction of American climate change policy did not disappear with the end of the negotiations in Kyoto (Fisher, 2004). Both the need to obtain congressional support for the Protocol and continuing international negotiations about the rules necessary for its implementation provided numerous opportunities for opponents to raise familiar objections. Members of Congress attacked the administration for failing to secure the meaningful participation of developing countries in efforts to reduce greenhouse gas emissions and expressed concern that the agreement would harm the American economy, powerful industrial groups argued that ratification would be a disaster for profits, workers, and consumers, and conservative groups continued to question the science of climate change. President Clinton sought to counter these arguments but a number of events conspired to undermine his authority. First, news of Clinton's affair with the White House intern Monica Lewinski emerged early in 1998 and the ensuing scandal consumed the attention of the administration making it difficult to push forward with any policy objectives. Second, the Republicans maintained their majority following the midterm elections of November 1998, leaving opponents of the Protocol in positions of institutional authority. Third, Clinton faced the normal ebbing of authority faced by virtually all presidents nearing the end of their term in office. These developments left the administration stumbling "to walk the fine line" necessary to bridge domestic concerns and international commitments following Kyoto. The United States signed the Kyoto Protocol at COP-4 in Buenos Aires in November 1998, but President Clinton decided not to submit it to the Senate for ratification after calculating that he lacked the necessary votes for approval. The "fine line" eventually became too difficult to walk and negotiations about the rules to implement Kyoto collapsed at COP-6 in The Hague in November 2000 when differences between the United States and European Union could not be resolved.

President Clinton issued a brief "Statement on the Kyoto Protocol on Climate Change" on 10 December 1997 in which he applauded the agreement as "environmentally strong and economically sound" and welcomed use of "the tools of the free market to tackle this difficult problem." Others were less convinced that the Protocol was "environmentally strong," "economically sound," or in the country's best interests. Some environmental groups argued that the compromise agreed at Kyoto failed to address the problem of climate change with sufficient urgency. The World Wildlife Fund, for example, described Kyoto as "a flawed agreement that will allow major polluters to continue emitting greenhouse gasses through loopholes" (Bennet, 1997). A number of prominent business leaders argued that the agreement would be costly and ineffective. Thomas Kuhn, president of the Edison Electric Institute, condemned the Protocol as "economic suicide" (Fialka, 1997). Congressional leaders warned that the prospects of ratification were remote. Senate Majority Leader Trent Lott (R. MS) told reporters that "I have made clear to the President personally that the Senate will not ratify a flawed climate change treaty" (Darwall, 2013, 178). Legislators from both parties demanded proof that the agreement would not harm the economy and criticised the administration for failing to secure the participation of developing countries in reducing greenhouse gas emissions.

President Clinton attempted to deflect criticisms of the Kyoto Protocol in his State of the Union Address on 27 January 1998. In the address he defended the need for action by claiming that: "... if we don't reduce the emission of greenhouse gases, at some point in the next century, we'll disrupt our climate and put our children and grandchildren at risk," before explaining that the Kyoto Protocol "committed our nation to reduce greenhouse gases through market forces, new technologies, energy efficiency." Finally, he dismissed suggestions that the agreement would damage the economy: "We have always found a way to clean the environment and grow the economy at the same time. And when it comes to global warming, we'll do the same." This was the second time that Clinton had mentioned climate change in a State of the Union Address but the substance of his message was different. In the address given on 4 February 1997 he had spoken briefly about the need to "reduce the greenhouse gases that challenge our health even as they change our climate," whereas in 1998 he made a defence of his administration's actions to address the issue and sought to rally support for the Kyoto Protocol. Few members of Congress were convinced by Clinton's rhetoric, however, leaving ratification of the treaty unlikely unless two tasks could be achieved. First, the administration needed to show that the commitment to greenhouse gas emissions contained in the Kyoto Protocol would not harm the American economy. Second, negotiations over the rules necessary to implement the agreement had to persuade developing countries to play a meaningful role in meeting emission reduction targets.

Janet Yellen, chair of the President's Council of Economic Advisors, sought to sooth congressional concerns about the economic costs associated with the Kyoto Protocol in testimony before a subcommittee of the House Commerce Committee in March 1998 (HCC, 1998). Yellen claimed that the cost of meeting the emission reduction targets contained in the Protocol would be negligible ($7–12 billion a year, or 0.1 per cent of GDP, between 2008 and 2012) if the administration's proposals for flexibility and joint implementation were accepted by other countries in the negotiations over how to implement the agreement. Opponents of Kyoto expressed disbelief at the administration's cost estimates and demanded to see the supporting evidence. When these were finally released after Republicans threatened a subpoena they showed that the estimates were based on assumptions about levels of emissions trading and the cost of carbon credits that neither environmental nor business groups regarded as realistic (Cass, 2006, 167). Other congressional committees launched their own hearings into the cost of compliance following Yellen's testimony giving prominence to industry claims that the costs of action outweighed any benefits (Keller, 2009, 108). One report by the Energy Information Administration produced for the Senate Committee on Science in October 1998 projected a reduction of 0.65–4.40 per cent in GDP in 2010 if the Kyoto Protocol was adopted (EIA, 1998). In short, the administration's efforts to assuage fears about the costs of Kyoto backfired badly. Rather than increase support for the Protocol, the cost estimates presented by Yellen reinforced scepticism among both Democratic and Republican legislators about Kyoto (Cass, 2006, 167).

The release of an EPA document in March 1998 that asserted that the agency had the authority to regulate carbon dioxide emissions from power stations under the Clean Air Act of 1990 fuelled Republican hostility towards the Kyoto Protocol (USEPA, 1998). Republicans feared that the document showed that the administration intended to implement the Protocol irrespective of whether the Senate ratified it or not. To counter such a possibility Senator John Ashcroft (R. MO) and Rep. Joe Knollenberg (R. MI) introduced bills (S. 2019 The Economic Growth and Sovereignty Protection Act; HR 3807 The American Economy Protection Act) to prohibit the use of federal funds to implement the Kyoto Protocol unless ratified by the Senate. Although no action was taken on either bill, Rep. Knollenberg secured initial support for an amendment to the Departments of Veterans Affairs and Housing and Urban Development, and

Independent Agencies Appropriations Act for Fiscal Year 1999 that achieved the same purpose. The amendment stated that no federal funds could be used "for the purpose of implementation, or in preparation for implementation, of the Kyoto Protocol." Interpreted literally the amendment prohibited any action by the federal government to improve energy efficiency, promote renewable energy, or discuss ways of implementing the Kyoto Protocol though Rep. Knollenberg insisted that the "main purpose" was to "ensure that existing regulatory authority is not misused to implement or to serve as a future basis for the implementation of the Kyoto Protocol in advance of its consideration and approval by the Senate of the United States" (Congressional Record, 29 July 1998, H6565). The amendment was eventually rejected after President Clinton threatened to veto the bill if it remained. Similar amendments were subsequently added to six appropriation bills for Fiscal 2000.

Negotiations over the rules needed to implement the Kyoto Protocol revealed an equally large gap between the United States and other countries. Two primary concerns underpinned the administration's approach to these negotiations. First, the administration knew that the Senate was unlikely to ratify the Protocol unless developing countries agreed to meaningful participation in the process of reducing greenhouse gas emissions. Not only did the Byrd-Hagel Resolution mandate such participation, but opponents of the agreement used the lack of restrictions on countries like China and India to dismiss the Protocol as unfair and ineffective. Second, the administration needed to make sure that the rules governing flexibility mechanisms (emissions trading, joint implementation, and sinks) gave the United States enough leeway to meet its greenhouse gas emissions targets at a reasonable cost. The administration hoped to buy sufficient reductions elsewhere to alleviate the need for drastic cuts in domestic greenhouse gas emissions. Both of these goals generated serious opposition. Developing countries continued to argue that any requirement to limit emissions would harm their economic growth while the European Union argued that action to cut domestic emissions should be the main policy response to climate change (Cass, 2007, 200–202, 205–6).

The administration had some success in negotiations at the COP-4 held in Buenos Aires in November 1998. At the beginning of the conference Argentina announced that it would agree to binding emissions targets and Kazakhstan soon followed suit (Darwall, 2013, 181). Hopes that other developing countries would also announce their intention of voluntarily agreeing emission targets, however, failed to bear fruit. Opposition from China and other developing countries forced the chair of the conference to take an item on voluntary commitments for developing countries off the agenda. American negotiators also secured agreement that detailed plans for the operation of flexible mechanisms should be developed in time for COP-6 scheduled for The Hague in November 2000. Success in persuading Argentina and Kazakhstan to commit to binding emissions targets and keeping options on flexible mechanisms alive prompted the United States to sign the Kyoto Protocol the day before COP-4 was scheduled to end. The decision was applauded by proponents of action to address climate change but couched in cautious terms by the administration (Cushman, 1998b). President Clinton failed to make a statement about the decision leaving Vice President Gore to make an announcement. In a written statement Gore warned that signing the Protocol placed no obligations on the United States and stated that: "We will not submit the Protocol for ratification without the meaningful participation of key developing countries in efforts to address climate change" (Gore, 1998). Senator Chuck Hagel (R.NE), co-sponsor of the Byrd-Hagel Resolution, taunted the administration: "If this treaty is good enough to sign, it's good enough to be submitted to the Senate for an open, honest debate" (Cushman, 1998b).

A key reason behind the decision to sign the Kyoto Protocol was the belief that such action would boost the administration's negotiating position in subsequent meetings. Just prior to the decision to sign, Senator Joe Lieberman (D. CT) had counselled Clinton that: "If we are not at the table, we cannot cajole or convince the developing nations to become part of the solution" (Cushman, 1998a). The idea that signing the Protocol would help to overcome the barriers that blocked agreement, however, was misplaced. Although the administration managed to secure agreement at COP-5 in Bonn in November 1999 that discussions should continue to flesh out the meaning of flexible mechanisms and resolve differences over the treatment of sinks in time for COP-6, efforts to persuade developing countries to play a more meaningful role in reducing greenhouse gas emissions failed to achieve any success. Faced with demands for action from the United States, developing countries began to insist on greater access to clean technology and additional financial resources to secure their cooperation (Royden, 2002, 458). Finding agreement on these complex issues proved impossible at COP-6 in The Hague. Developing countries refused to budge on the question of greater participation, and disagreement between the United States and the European Union over flexible mechanism and the treatment of sinks eventually led to the collapse of the talks.

Conclusion

In his last State of the Union Address delivered on 27 January 2000, President Clinton devoted three paragraphs to climate change. He identified climate change as "the greatest environmental challenge of the new century" and warned that "if we fail to reduce the emission of greenhouse gases, deadly heat waves and droughts will become more frequent, coastal areas will flood, and economies will be disrupted. This is going to happen, unless we act." He also dismissed claims that cutting greenhouse gas emissions would cause economic harm with the claim that "new technologies make it possible to cut harmful emissions and provide even more growth." The address testifies to the transformation in Clinton's attitude towards climate change that occurred during his period in office. In his early years Clinton had focused primarily on economic issues and treated climate change with considerable caution. An improving economy which provided opportunities to pursue other policy objectives, growing scientific certainty about the cause and consequence of climate change, and the demands of the process created by the Rio Treaty gradually led Clinton to abandon this caution. The nature of the problem, policy, and political streams at the time, however, meant that radical policy action proved impossible. Sufficient uncertainty about the science enabled opponents to continue to question the need for action, no consensus existed on the best way to address the problem, the public lacked engagement with the issue, climate sceptics dominated Congress, and international actors had vastly differing views about how to operationalise the Kyoto Protocol. Clinton had the authority to put climate change on the political agenda but lacked the power to change policy in a radical way.

Chapter 4
Scepticism, Neglect, and Obstruction

The election of George W. Bush in November 2000 gave a boost to opponents of government action to address climate change. President Bush had a background in the oil industry and during the election campaign had expressed scepticism about the science of climate change, asserted that tackling the problem would harm the American economy, and cast doubt on the value of the Kyoto Protocol. Although the disputed outcome of the presidential election muted talk of mandates, Bush had no compunction about using his executive authority to shape American policy on climate change. His first six years in office were marked by efforts to abrogate international obligations, weaken or obstruct domestic commitments to dealing with the problem, and undermine climate change science. The results of congressional elections furthered this programme of scepticism, neglect, and obstruction (Harris, 2009). Republicans maintained control of the House of Representatives until 2006 and had a majority in the Senate from January 2001 to June 2001, and from 2002 to 2006. Compliant Republican majorities limited opportunities to challenge presidential initiatives and meant that opponents of action to address climate change occupied positions of institutional power. Republican control of Congress left state governments as the main sources of opposition to administration policy on climate change during this period. Various state governments launched important legal challenges to decisions made by the administration and also initiated their own climate change policies. Democratic victories in the 2006 midterm elections, however, altered the politics of climate change in Washington, DC. Advocates of strong action to tackle climate change assumed positions of power in Congress and used their authority to investigate the administration's actions and launch their own initiatives. Faced with challenges to act President Bush eventually softened slightly his approach to climate change and proposed some modest changes in policy.

Intense struggles over problem definition, issue framing, and potential solutions are evident during the Bush Administration. Debates about the "problem" dominated policymaking in the early years of the decade with sceptics often dismissing the idea that a problem existed while others warned of impending catastrophe. Battles over "framing" raged as a result (Fletcher, 2009; Nisbet, 2009; Vezirgiannidou, 2013). President Bush initially sought to frame the issue in terms of scientific uncertainty, economic cost, and international competitiveness, but changed tack in later years to talk about the long-term nature of the problem as the weight of scientific evidence made it difficult to sustain claims about an uncertain science. On the other hand, advocates of action to tackle climate change sought to frame the issue as an urgent problem that posed a real threat to the country, offered opportunities to build a "green economy" that would produce benefits rather than costs, and began to talk about the problem as a social issue. Debates about potential solutions also raged. President Bush placed great faith in technological innovation as a means to tackle the problem, rejected the imposition of mandatory limits on greenhouse gas emissions, and viewed multilateral approaches to addressing the problem with suspicion. Opponents stressed the need for mandatory emission limits for greenhouse gases, started exploring the idea of a "cap-and-trade" system as a means of enforcing such limits, and urged the United States to play a meaningful role in multilateral negotiations. In short, considerable stirring of the "policy primeval soup" occurred during this period leaving little settled. Changes in the "political" stream, particularly electoral outcomes, privileged certain voices in the debates about the problem and potential solutions at different times. Sceptics dominated the policy process in the early years of the Bush Administration but found their power increasingly challenged following the 2006 midterm elections.

The 2000 Election

The 2000 presidential election offered voters a choice between two candidates with different views on climate change. Vice President Al Gore had established a reputation as an environmentalist with the publication of *Earth in the Balance* in 1992 and had played a significant role in brokering the Kyoto Protocol, while Governor George Bush had a record as a supporter of the oil industry and a reputation as a climate change sceptic. Gore made a strategic decision early in the campaign, however, not to emphasise his environmental credentials in an effort to appeal to voters in swing states (Mencimer, 2002). Rather than seeking to persuade voters of the need to take action to reduce greenhouse gas emissions, he chose to speak about technological advances and economic growth in an effort to forestall Republican claims that he had a radical environmental agenda that would damage the American economy. Gore calculated that pushing an aggressive environmental agenda would harm his electoral prospects. Differences between the two candidates on climate change emerged late in the campaign, but Al Gore's early caution not only allowed George Bush to frame the issue to his advantage but also prompted Ralph Nader to enter the campaign as a standard bearer for the environmental movement. Nader, the Green Party candidate, believed that he offered voters concerned about the environment an alternative to Gore (Nader, 2002).

Climate change rarely figured in the early stages of the 2000 presidential election nomination campaigns. Vice President Al Gore promised that: "I will address the international challenge of global warming with new technologies that create more jobs, and make our economy even stronger" when announcing his candidacy on 16 June 1999, but did not make any major public statements on the issue until a year later when he sought to counter challenges that environmental protection would lead to increased bureaucracy and economic costs. In a speech given in

New York on 13 June 2000 he argued that "There will be no new bureaucracies; no new agencies or organizations, because not only is the era of big government over, the era of old government is over, too ... through the power of free markets, through good old-fashioned American ingenuity, we will dramatically reduce pollution and reverse the tide of global warming—while creating more jobs, not fewer jobs, for our people." Republicans paid even less attention to climate change in the nomination campaign. Only Senator Orrin Hatch (R. UT) raised the issue when he declared in the Republican Presidential Debate held in Des Moines, Iowa, on 13 December 1999 that: "I don't agree with environmental extremism that would make us uncompetitive with the rest of the world. One of the first things I said I would do [as President] is revoke the Kyoto Accords. The Kyoto Accords place environmental extreme requirements on the United States, and nobody else. And in the final analysis, there's no real reason—scientific or otherwise—why that should occur."

The lack of attention given to climate change in the early stages of the 2000 presidential election reflected Vice President Gore's calculation that an aggressive emphasis on environment issues would harm his electoral prospects. Gore and his advisors believed that making the environment a central focus of the election would alienate potential business supporters, damage his centralist "New Democrat" image, and bring little reward as the public regarded the issue as a low priority (Mencimer, 2002). A Gallup Poll conducted on 22 April 1999 revealed that public concern about the environment had waned over the previous decade and global warming did not figure high as a priority among environmental problems (Gallup, 1999). While 68 per cent of those polled stated that they worried "a great deal" about the pollution of drinking water only 34 per cent claimed that they worried "a great deal" about global warming. Only acid rain came below global warming in the list of nine environmental problems that respondents were asked about. A Gallup Poll conducted almost exactly a year later revealed a similar picture (Gallup, 2000a). The poll revealed that Americans were most concerned about education, health care, and crime with environmental protection ranking eighth in a list of major concerns. Global warming continued to rank low among a list of 11 environmental problems. Although 40 per cent of those polled stated that they worried "a great deal" about global warming, the problem ranked a long way below drinking water quality, the pollution of rivers and lakes, and toxic waste sites. Only acid rain was seen as a less pressing problem. The two polls also revealed partisan differences with Democrats identifying environmental protection as a higher priority than Republicans.

Publication of the Party Platforms in the summer of 2000 revealed a divide between the two parties over climate change despite Vice President Gore's downplaying of the issue. The Republican Party Platform published on 31 July 2000 included a short statement on global warming that claimed the issue was "complex and contentious," questioned the science that underpinned the Kyoto Protocol, argued that the "agreement would be ineffective and unfair," and called for more research on the issue. The Platform stated that "A Republican president will work with business and with other nations to reduce harmful emissions through new technology without compromising America's sovereignty or competitiveness— and without forcing Americans to walk to work." The Democratic Party Platform published on 14 August 2000 contained a long section on climate change that identified global warming as a problem that could lead to "much of Florida and Louisiana under water … floods, droughts, diseases and pests. Crop failures and famine …" and declared that "[w]e have to do what's right for our Earth because it's the moral thing to do." Proposals for investment in mass transit systems, action to promote the development and use of clean energy technology, and ratification of the Kyoto Protocol were advanced to tackle the problem. Promises that "there will be no new bureaucracies, no new agencies, no new organizations" and that we will make "sure that all nations of the world participate in this effort" sought to head off Republican charges that action would lead to big government, and place the United States at an economic disadvantage if other countries failed to act.

The differences between the two candidates on environmental issues bubbled to the surface when Governor Bush made a speech on energy in Saginaw, Michigan, on 29 September 2000. In a speech to auto workers Bush called for more domestic fuel production, opening up the Arctic National Wildlife Refuge, incentives for developing alternative energy sources, and in a surprising move, argued that power stations should be required to reduce emissions of pollutants, including carbon dioxide (Bruni, 2000). Bush's call for mandatory reductions in carbon dioxide emissions was hailed by environmentalists but astonished Republicans who had previously rejected such measures on the grounds of cost. Conservative newspaper columnist Robert Novak later claimed that the promise to regulate carbon dioxide emissions was simply a mistake made in the heat of the campaign (Jalonick, 2004). Vice President Gore responded immediately in a speech to the National Audobon Society in Chevy Chase, Maryland, in which he took issue with Bush's call for more drilling. Gore declared "I believe we don't have to degrade our environment in order to ensure our energy future," and called for greater investment in clean energy technology and mass transit systems. His caution about advancing an aggressive environmental agenda, however, limited the force of Gore's counter attack. He stuck to familiar themes and did not respond at all to Bush's promise about regulating carbon dioxide emissions. Caution about making an issue of climate change in the campaign continued to shape Gore's electoral strategy.

Exchanges between the two candidates over climate change were few and far between during the campaign (McCright and Dunlap, 2011). The Presidential Debate held in Winston-Salem on 11 October 2000 proved the main occasion when Vice President Gore and Governor Bush engaged with each other on the issue. Bush stated that global warming "is an issue that we need to take very seriously," but raised questions about the certainty of the science, cast doubt on whether enough was known about the problem to identify solutions, claimed that collaboration at state and local levels rather than "command-and-control out of Washington DC" should be the way forward, and made plain that "I'm not going to let the United States carry the burden for cleaning up the world's air." Gore argued that a scientific consensus on global warming existed, agreed that command-and-control did not offer the best way to deal with the problem, and claimed that "we can create millions of good new jobs by being the first into the market" for clean energy technologies. Both candidates essentially articulated policy positions that had not changed since the beginning of the campaign and provided voters with a real choice. Neither gave the impression that climate change was a priority.

Evidence that the policy differences between Governor Bush and Vice President Gore over the environment in general, and climate change in particular, played a significant role in determining the outcome of the disputed 2000 presidential election is mixed. On one hand, opinion polls suggest that the environment ranked ninth, far behind education, the economy, and health care in a list of issues that voters identified as influencing their vote (Gallup, 2000b). Global warming also continued to lag behind all other environmental concerns except acid rain as a priority among the public. The evidence seems to suggest that few people cast their votes based on the candidates' position on the issue. On the other hand Gore's failure to campaign aggressively on the environment may have led to voters deserting him for Ralph Nader (Mencimer, 2002; Jalonick, 2004). National exit polls suggest that about half of Nader's votes would have gone to Gore which would probably won him the election. Nader gained more than 95,000 votes, for example, in the disputed state of Florida. In the highly unusual circumstances of the 2000 election Gore's lukewarm approach to the environment may have cost him victory.

The result of the election and those for the US Congress changed the "political stream" sufficiently to have an impact on policy (Kingdon, 2011). First, a Republican administration sceptical about the need for action and with close links to the fossil fuel industry replaced a Democratic administration sympathetic to the idea that climate change was a problem that needed to be addressed. Not only did President George Bush and Vice President Dick Cheney have a background in the oil industry, but the energy sector also contributed large sums of money to the Republican campaign (Beder, 2002). The Centre for Responsive Politics claims that Political Action Committees (PACs) connected to the energy and natural resources sector, for example, contributed just over $12 million to Republican candidates in the 2000 presidential and congressional elections compared to $5 million to Democratic candidates. The election both changed control of the Executive Branch and gave groups opposed to limits on greenhouse gas emissions privileged access to policymakers. Second, the results of the congressional elections left Republicans in control of the House of Representatives and the Senate. Although the decision of Senator James Jeffords (R. VT) to become an Independent in June 2001 handed control of the Senate to the Democrats, this change of party control proved short-lasting as the Republicans regained control following the 2002 midterm elections. Republican control of Congress from 2002 until 2006 restricted opportunities for challenges to presidential action and the development of alternative policy proposals. Finally, opinion polls suggested an ambivalence in the public mood for action. Americans identified global warming as a problem but its saliency ranked very low.

Repudiating Kyoto

The first sign that the 2000 election presaged a shift in climate change policy came when President Bush began to announce his appointments to posts within the new administration. People with backgrounds in the energy, aluminium and automobile industries dominated these appointments giving industry groups a privileged voice in the deliberations and decision-making venues of government (Lisowski, 2002; Byrne et al., 2007). Chief of Staff Andrew Card had worked as chief lobbyist for General Motors and the American Automobile Manufacturers Association; Commerce Secretary Donald Evans had close links with an oil and gas company based in Denver; Treasury Secretary Paul O'Neil had served as chairman of Alcoa; National Security Advisor Condoleeza Rice had served on Chevron's board of directors; and Energy Secretary Spencer Abraham had received large contributions from the automobile industry when campaigning for the Senate in 2000. Others had records of opposition to federal action to protect the environment. Interior Secretary Gale Norton had long been a champion of using "property rights" to restrict the regulatory power of the federal government, and her appointment was strongly opposed by environmental groups who feared that she would allow energy companies to drill in Alaska's National Wildlife Refuge (Jehl, 2000, 2001a). James Connaughton, chair of the Council on Environmental Quality, had acted as a lawyer for utility companies in lawsuits against the Environmental Protection Agency (EPA). The appointment of Christine Todd Whitman as EPA administrator provided a notable exception to this catalogue of appointments with connections to major industries. Whitman had earned plaudits from many environmental groups for her actions as governor of New Jersey to improve air quality and clean up beaches, but quickly found herself isolated in the new administration (Whitman, 2005).

Evidence that President Bush intended to alter the direction of climate change policy occurred early in 2001 when the new administration requested that the resumption of the 6th Conference of the Parties (COP-6) scheduled for May 2001 be postponed until July 2001 (Cass, 2006). Officials at the State Department explained that this would enable the administration to re-evaluate America's climate policy. Bush provided a clear statement of the scope of this re-evaluation in a letter to four Republican senators dated 13 March 2001 that reneged upon his campaign pledge to regulate carbon dioxide emissions from power stations, and reaffirmed his opposition to the Kyoto Protocol (Jehl and Revkin, 2001). A major energy crisis in California provided the cover to abandon the campaign promise. Bush stated that the government would not impose mandatory emissions reductions for carbon dioxide from power stations because it would lead to a rise in electricity prices at a time of energy shortages. The announcement undermined EPA Administrator Whitman who a few weeks earlier had given an interview on CNN in which she stated "George Bush was very clear during the course of the campaign that he believed in a multi-pollutant strategy, and that includes CO_2, and I have spoken to that" (CNN, 2001). In his letter Bush also gave notice of his misgivings about Kyoto. He pointed out that the agreement "exempts 80 per cent of the world, including major population centers such as China and India, from compliance," argued that it "would cause serious harm to the US economy," and suggested that it was based on "the incomplete state of scientific knowledge of the causes of, and solutions to, global climate change." Two weeks later EPA Administrator Whitman formally announced that: "We have no interest in implementing this treaty" (Jehl, 2001b). Neither decision registered particularly highly with the public. Despite widespread media coverage of both decisions only 28 per cent of those polled reported knowledge of the reversal of Bush's campaign pledge about carbon dioxide emissions and only 20 per cent expressed awareness of the Kyoto decision (PEW, 2001). European leaders, on the other hand, denounced the decision to withdraw as "arrogant," "irresponsible," and "sabotage" (Andrews, 2001). They noted that US withdrawal from the Kyoto Protocol meant that the world's largest producer of carbon dioxide was no longer party, even at a symbolic level, to an international agreement to reduce greenhouse gas emissions.

Ari Fleischer, the White House press secretary, initially employed sophistry to defend the administration's Kyoto decision. In a press briefing given on 28 March 2001 he argued that the United States could not withdraw from a treaty that had not been ratified by the US Senate "because there is no treaty in effect." He then proceeded to reiterate that President Bush believed that the Kyoto Protocol was flawed because it "exempts the developing nations around the world, and it is not in the United States' best interests." Fleischer claimed that Bush recognised that global warming was a serious problem and had ordered a Cabinet-level review to determine the best way forward. He stated that: "It's a question of what we can do based on sound science and a balanced approach as a nation to take action against global warming." The following day President Bush explained his decision on two separate occasions. In an exchange with reporters prior to meeting with Chancellor Schroeder of Germany he argued that "the idea of placing caps on CO_2 does not make economic sense" with the economy slowing and an energy crisis brewing. "And while I worry about emissions," he continued, "I'm also worried that people may not be finding jobs in America." He pledged to work with allies like Germany to tackle climate change but warned that decisions would be based on "what's in the interest of our country, first and foremost." In a later news conference he returned to these themes when he stated that: "we will not do anything that harms our economy, because first things first are the people who live in America. That's my priority. And I'm worried about the economy. I'm worried about the lack of an energy policy. I'm worried about rolling blackouts in California."

The reasons given by President Bush for repudiating the Kyoto Protocol echoed comments made on the campaign trail. Bush had consistently talked about a "flawed" treaty that exempted major producers of greenhouse gases from regulation, made claims about the uncertainty of climate science, and argued that the agreement would harm the American economy. Some of these concerns might have been assuaged through further international negotiations, but Bush had little interest in discussions about a treaty that effectively committed the country to a policy of fossil fuel conservation at a time when his administration was planning a national energy policy based on increasing fossil fuel supply (Lisowski, 2002). President Bush had established an energy task force in January 2001, chaired by Vice President Cheney, with the goal of expanding oil and gas production as a means of solving the country's energy supply problems. The task force's report, published in May 2001, rejected conservation as a means of dealing with the country's energy problems, recommended drilling for oil and gas in the Arctic National Wildlife Refuge and other environmentally sensitive areas to boost the supply of fossil fuels, and advocated greater use of nuclear energy. Evidence emerged later that Cheney and his officials had relied heavily upon their contacts in the energy industry when drafting the report (Abramowitz and Mufson, 2007; Byrne et al., 2007).

One month after release of the energy task force's report the interim findings of the Cabinet-level review into climate change ordered by President Bush in March 2001 were made public. In Remarks on Global Climate Change made on 11 June 2001 Bush outlined the review's findings. He began by discussing a report on climate science that had been conducted by the National Academy of Sciences as part of the Cabinet-level review (NAS, 2001). Although the report was broadly consistent with the Third Assessment Report of the IPCC, President Bush emphasised the parts that noted a lack of knowledge or debates among scientists in keeping with Republican efforts to frame climate change in terms of scientific uncertainty (McCright and Dunlap, 2010). Bush stated that the NAS report confirmed that the surface temperature of the Earth is warming, acknowledged that the "National Academy of Sciences indicates that the increase [in greenhouse gases] is due in large part to human activity," but stressed that "the Academy's report tells us that we do not know how much effect natural fluctuations in climate may have on warning." As well as casting doubt on the cause of climate change Bush also questioned the impact of global warming. "No one can say with any certainty what constitutes a dangerous level of warming and, therefore, what level must be avoided," he argued. Bush urged caution rather than precaution in the face of this uncertainty. He stated that: "The policy challenge is to act in a serious and sensible way, given the limits of our knowledge. While scientific uncertainties remain, we can begin to address the factors that contribute to climate change."

President Bush's emphasis on "the limits of our knowledge" and "scientific uncertainties" followed the advice given by Republican pollster Frank Luntz in a memo about how to deal with the issue (Burkeman, 2003). Luntz had counselled Bush to stress "your commitment to sound science," argue that "the scientific debate "about global warming remains open," and call for more research. This would not only help them to appear reasonable, but also help to confuse the public with competing claims about climate change. Luntz's advice on how to deal with climate change can also been seen in the policy initiatives announced by President Bush in his 11 June 2001 Remarks on Global Climate Change. Bush began by offering a conciliatory statement that "America's unwillingness to embrace a flawed treaty should not be read by our friends and allies as any abdication of responsibility. To the contrary, my administration is committed to a leadership role on the issue of climate change." He then proceeded to outline "a number of initial steps" recommended by the Cabinet-level review. First, he established a US Climate Change Research Initiative to study areas of uncertainty because "we need to know a lot more" about the causes and consequences of climate change. Second, he created a National Climate Change Technology Initiative to promote the development of technology to monitor and reduce greenhouse gases. Third, he pointed to support for clean energy technologies contained in the energy plan produced by Vice President Cheney as evidence of a commitment to find long-term alternatives to fossil fuels. And finally, he declared that "we will work with other countries to monitor, measure, and mitigate emissions." This commitment to action, however, was tempered by a familiar reminder that "We must always act to ensure continued economic growth and prosperity for our citizens and for citizens throughout the world. We should pursue market-based incentives and spur technological innovation. And finally, our approaches must be based on global participation."

Environmental groups dismissed the "initial steps" outlined by President Bush in his 11 June 2001 speech as an effort to evade the issue by promising new scientific initiatives without detailing how much money would be provided for such studies or specifying a timetable for their completion (Sanger, 2001). European leaders also dismissed Bush's initiative. EU Environment Commissioner Margot Wallström stated that: "We think it is time to move on from analysing the issues towards action. I am worried that his speech is short of action to actually reduce emissions" (EU, 2001). The extent of the gap between President Bush and European leaders on climate change became even more apparent during a US-EU Summit in Götenburg, Sweden on 13–14 June 2001. In a press conference held on 14 June 2001 Swedish Prime Minister Goran Persson noted that: "We don't agree upon how we regard the Kyoto Protocol—so to say we agree to disagree about substance." Bush responded that: "… we don't agree on the Kyoto treaty, but we do agree that climate change is a serious issue and we must work together. We agree that climate change requires a global response and agree to intensify cooperation on science and technology." The official statement of the Götenburg Summit recognised that climate change is "a pressing issue," noted the disagreement on Kyoto, but pledged cooperation between the US and EU to address the issue. One month later President Bush gave a report on progress that had been made since announcing his "initial steps." In a Statement on Climate Change Review Initiatives made on 13 July 2001 he spoke about the creation of a new interagency group to develop a Federal research plan, new projects to study carbon sequestration, a range of bilateral initiatives with other countries, and a commitment to "participate constructively" in international discussions on climate change. Environmental groups remained unimpressed. "What these initiatives lack is anything that would actually reduce global warming pollution" stated David Hawkins, director of the NRDC's Climate Center, "The announcement is pure cotton candy—a spoonful of sugar spun into a ball of fluff" (NRDC, 2001).

President Bush's report on the progress that had been made since announcing his "initial steps" came just days before the rescheduled COP-6 began in Bonn, Germany, on 19 July 2001. In her opening statement Paula J. Dobriansky, under secretary of state for democracy and global affairs, told the conference that the United States "intends to be a constructive and active party to the UN Framework Convention on Climate Change" and declared that "though the United States will not ratify the Kyoto Protocol, we will not abdicate our responsibilities" (Dobriansky, 2001a). Some delegates booed Dobriansky as she made her remarks and the American delegation played no substantial role in the negotiations that led to important revisions to the Kyoto Protoco (Payne and Payne, 2014, 360). The withdrawal of the United States from discussions about the Protocol strengthened the position of Australia, Canada, Japan, and Russia in negotiations, however, and led to a watering down of emission targets and agreement that credit would be given for maintaining large forests as "carbon sinks." Critics charged that the Bonn Accords significantly weakened the emission reduction targets contained within the Kyoto Protocol, but European leaders hailed the result as evidence that progress could be made without the participation of the United States (Kahn, 2003; Cass, 2006). American isolation continued when negotiations began at COP-7 in Marrakech, Morocco, in November 2001 to resolve remaining issues. President Bush's decision to withdraw from the Kyoto Protocol effectively sidelined the American delegation and Dobriansky had little to say beyond repeating the familiar mantras of the administration. In her closing statement on 9 November 2001 she told the conference that the "Protocol is not sound policy" and the agreed "emission targets are not scientifically based or environmentally effective" (Dobriansky, 2001b).

President Bush announced a Global Climate Change Initiative on 14 February 2002 that offered a different approach to dealing with the issue than that agreed in Bonn and Marrakech (Selin and VanDeveer, 2007). The Initiative called for an 18 per cent reduction in greenhouse gas *intensity* relative to GDP by 2012 rather than an *absolute* reduction in emissions, relied upon *voluntary* rather than *mandatory* action to achieve this reduction, and proposed further research into climate change (Bang et al., 2005). Bush stated that establishing a greenhouse gas intensity target "is the common sense way to measure progress." He argued that "economic growth and environmental protection go hand in hand," claimed that "economic growth is the solution, not the problem," because growth is "what pays for investment in clean technologies, increased conservation, and energy efficiencies," and recommended that developing countries such as China and India adopt the same approach to help them meet "some of the shared obligations" in addressing climate change. Environmentalists attacked the Initiative on two main grounds. First, the target of an 18 per cent improvement in greenhouse gas intensity over 10 years would allow an increase in actual emissions. Greenhouse gas intensity is a ratio of emissions to economic output, and can fall even if actual emissions are rising. Studies have shown that during the 1980s and 1990s greenhouse gas intensity fell in the United States despite a continuous increase in emissions (Van Vuuren et al., 2002; Matsuo, 2002). Second, achieving emission targets through voluntary action has a poor track record. For example, the United States, like several other countries, failed to meet the voluntary target of reducing emissions of greenhouse gases to 1990 levels by 2000 that it agreed at the Rio Conference on Environment and Development in 1992.

The language used by President Bush in his Remarks announcing the Global Climate Change Initiative on 14 February 2002 hinted at a softening in the administration's attitude towards climate change. Bush noted that "we must address the issue of global climate change," declared that "wise action now is an insurance policy against future risks," and reaffirmed "America's commitment to the United Nations Framework Convention and its central goal, to stabilize atmospheric greenhouse gas concentrations at a level that will prevent dangerous human interference with the climate." Familiar statements about scientific uncertainty, economic growth, and the need for all countries to take action, however, sat alongside these murmurs of action, precaution, and engagement. Bush talked about "scientific uncertainties," claimed that "the answers are less certain," and stressed that decisions must be based on "sound science." He declared that "our nation must have economic growth, growth to create opportunity, growth to create a higher quality of life for our citizens," and reiterated that "developing nations such as China and India already account for a majority of the world's greenhouse emissions, and it would be irresponsible to dissolve then from shouldering some of the shared obligations." Rather than signalling a change in the administration's approach to climate change, Bush's rhetoric articulated positions consistent with those expressed on the campaign trail in 2000. Economic growth remained the priority, the United States would not bear an unfair burden in addressing the problem, and care needed to be taken in interpreting the science.

The package of policies contained in the Global Climate Change Initiative reflected the fact that little had changed in the administration's thinking about the problem. Mandatory limits on greenhouse gas emissions were rejected in favour of a call for voluntary action to meet a target for a reduction in greenhouse gas intensity accompanied by proposals for further research. The Initiative set out a number of objectives for promoting new technology, transportation programmes, and carbon sequestration, but failed to detail how the federal government would achieve these objectives (Carlarne, 2010, 40–44). Faced with international and domestic demands for action, President Bush responded with a mix of proposals "designed mainly to give the appearance of a credible response" (Victor, 2004, 130). His speech made no mention of the need for hard choices, appealed to an American faith that technological advances would obviate the need for changes in behaviour, and promised that increased affluence would solve all problems (Cannon and Riehl, 2004).

A further indication that little had changed in the administration's attitude towards climate change came in June 2002 when the EPA published the *US Climate Action Report—2002*. This Third National Communication to the UNFCCC acknowledged that human action was a major cause of global warming, and specified the likely impact of climate change in the United States, but failed to propose any major shift in policy (Revkin, 2002a). The report did not recommend reductions in greenhouse gas emissions but called instead for action to adapt to climatic changes that might occur in the future. Environmentalists criticised the report. Patrick Mozza of Earth Island stated that: "What this report says in essence is global warming is coming, it's going to be serious, but just lay back and take it" (Clarke, 2002). Even the suggestion that human activity was a major cause of climate change that would adversely impact upon the United States, however, proved anathema to President Bush. In comments to reporters on 4 June 2002 he distanced himself from the report by describing it as something "put out by the bureaucracy" (Seelye, 2002). Later that day Ari Fleischer, the White House press secretary, highlighted areas of scientific uncertainty in the report to undermine its value. At a press briefing he stated "there is 'considerable uncertainty'—that's in the recent report—relating to the science of climate change. This report submitted to the United Nations also recognizes that any 'definitive prediction of potential outcomes is not yet feasible' and that 'one of the weakest links in our knowledge is the connection between global and regional predictions of climate change.'"

Over the next few months evidence began to emerge that the Bush Administration's efforts to frame climate change in terms of scientific uncertainty went beyond the use of selective quotations that suggested a lack of consensus, but also encompassed a systematic attempt to interfere in the work of climate scientists within the federal government. Changing or suppressing scientific results formed a key tactic (Mooney, 2006; Shulman, 2008; Shapiro, 2009). In September 2002 the administration removed an entire section on climate change from the EPA's annual air pollution report, and followed this nine months later by trying to force the EPA to remove data on global warming from its "Report on the Environment" (Revkin, 2002b; Revkin and Seelye, 2003). Subsequent surveys of federal scientists by the Union of Concerned Scientists (UCS) revealed that almost half of respondents had been pressured to remove words such as "climate change" or "global warming" from documents, and a similar number believed that officials had made inappropriate changes to their work (Mooney, 2008). Disclosure of the extent of the administration's interference in the work of climate scientists eventually forced the resignation of White House Council on Environmental Quality chief of staff Philip Cooney in June 2005 (Revkin, 2005). Cooney, who had previously worked in the oil industry, had played a major role in editing scientific reports to give an impression of scientific uncertainty and a lack of consensus on climate change. Following his resignation Cooney accepted a job with ExxonMobil (Kolbert, 2006).

The Bush Administration made an effort to sell the approach to dealing with climate change announced in the Global Climate Change Initiative at the COP-8 in New Delhi, India, in October 2002. Noting that the conference was the first since President Bush had announced the Initiative, US Special Envoy to the UNFCCC Harlan L. Watson held a press briefing on 24 October 2002 "to inform other Parties of our climate change policy and to explain it in more detail" (Watson, 2002a).

He reiterated that the United States will not ratify the Kyoto Protocol and would not sign any successor agreement when the Protocol expired in 2012. In his statement and answers to questions he stressed the need for an approach that linked action on climate change to economic growth. "Economic growth is an absolute necessity, certainly for developing nations," he claimed, because "the world simply has to become richer, if nothing else, so that we can address the basic needs of the public on a global basis, and to have the resources to invest in the new technologies that we are going to need to address climate change in the long term." He returned to this theme in his final remarks to the conference on 25 October 2002 when he argued that climate change needed to be addressed in the broader context of sustainable development (Watson, 2002b). In an effort to persuade developing countries that the American approach to dealing with climate change better suited their interests than that found in the Kyoto Protocol, Watson declared that "We must recognize that it would be unfair—indeed counterproductive—to condemn developing nations to slow growth or no growth by insisting that they take on impractical and unrealistic greenhouse gas emissions."

Hopes that the Global Climate Change Initiative would form the basis of an alternative regulatory regime to that established by the Kyoto Protocol proved unfounded. Both developed and developing countries continued to ratify the Protocol and the treaty eventually came into force following Russia's ratification in November 2004. Efforts to promote the administration's approach to climate change, however, continued with the negotiation of a number of bilateral and multilateral agreements with other countries and regional organisations. In a presentation to a seminar organised by the UNFCC in Bonn, Germany, in May 2005 Harlan Watson told participants that the United States had entered into agreements with 14 countries and regional organisations responsible for nearly 80 per cent of world greenhouse gas emissions (Watson, 2005). These agreements covered areas such as research into climate change, and the development of clean energy technologies and methods of carbon capture. Two months after Watson's presentation the United States created the Asia Pacific Partnership on Clean Development and Climate with Australia, Japan, China, Korea and India to improve cooperation on the development of new technologies (Oliver, 2005). None of the agreements sought to regulate greenhouse gas emissions.

The Bush Administration's claims that more research was needed to resolve uncertainties in the science of climate change before taking action on greenhouse gas emissions became more difficult to sustain as scientific knowledge improved. Initiatives to promote research, in effect, proved a double-edged sword for the administration. On the one hand they served as a delaying tactic. Doing nothing could be justified by a need to make sure that the problem was properly understood. On the other hand they eventually produced results that improved knowledge about the problem and strengthened the case for action. Efforts to censor the findings of government scientists proved possible in the short-term but could not be maintained and provided opportunities for opponents to make accusations about subverting scientific integrity. Even less could be done about research conducted elsewhere that improved knowledge of the causes and consequences of climate change. The weight of these findings finally forced the administration to acknowledge the connection between human activity and climate change. In "Our Changing Planet," a report by the Climate Change Science Program and Subcommittee on Global Change Research transmitted to Congress on 26 August 2004, conceded that greenhouse gas emissions provided the only explanation for global warming over the previous few decades (Rahm, 2010). No change in policy followed this acceptance of the science, however, as administration officials continued to argue that the economic cost of mandatory emissions limits were too high and would have no effect if developing countries were not required to take similar action.

The policy vacuum left by the Bush Administration's failure to address climate change in the early 2000s galvanised a number of state and local governments to take action to address the problem (see Rabe, 2004; 2007; Byrne et al., 2007; Lutsey and Sperling, 2008; Selin and VanDeveer, 2009; Posner, 2010). Three particular initiatives stand out among a range of state actions. First, California passed a Global Warming Solutions Act in 2006 that provided the basis for a cap-and-trade programme in the state beginning in 2013. Only the European Union's Emission Trading System covered more emissions than California's programme. Second, states began to develop renewable portfolio standards (RPSs) that required utility companies to generate a specific amount of electricity from renewable sources (Rabe, 2004). By January 2009, 33 states had RPSs. Third, states began forming regional pacts, often involving Canadian provinces, to reduce emissions of greenhouse gases. In 2005 seven north-eastern states signed the Regional Greenhouse Gas Initiative (RGGI) to cut emissions to 1990 levels by 2010 and to achieve a level 10 per cent below 1990 levels by 2020. In February 2007 the governors of Arizona, California, New Mexico, Oregon, and Washington state formed the Western Climate Initiative (WCI) to tackle global warming. Utah, Montana, and the Canadian provinces of British Columbia, Manitoba, Quebec, and Ontario subsequently joined the WCI. A further regional pact was formed in November 2007 when nine Midwestern governors and the premier of Manitoba signed the Midwest GHG Reduction Accord to reduce carbon emissions and set up a trading system to meet reduction targets (Broder, 2007). Many local governments also used their power over building regulations, land use, public transport, and recycling to address climate change during this period. Several studies have suggested that the impact of these state and local initiatives to reduce greenhouse gas emissions would be substantial if implemented fully (Byrne et al., 2007; Lutsey and Sperling, 2008).

The 2004 Election

Climate change barely figured in the 2004 presidential election. Neither President Bush nor Senator John F. Kerry made a major speech on climate change and references to the issue in the party platforms were brief and vague. The

Democratic Party Platform published on 26 July 2004 noted that "... even though overwhelming scientific evidence shows that global climate change is a scientific fact, this administration has rewritten government reports to hide that finding," but lacked details about what the Democrats would do to address the problem. The Platform asserted vaguely that "[we] will address the challenge of climate change with the seriousness of purpose that this great challenge demands ... We must restore American leadership on this issue." The Republican Party Platform published on 30 August 2004 contained none of the overt scepticism found in the 2000 Platform but framed climate change as a long term issue. The Platform stated that "Republicans are committed to meeting the challenge of long-term global climate change by relying on markets and new technologies to improve energy efficiency" and reiterated that "Our President and our Party strongly oppose the Kyoto Protocol and similar mandatory carbon emission controls that harm economic growth and destroy American jobs." Climate change emerged briefly as an issue in the Presidential Debate held in St Louis, Missouri, on 8 October 2004. In an exchange Senator Kerry attempted to attack the President over his failure to accept the findings of climate change scientists and declared that "I'm going to be a President who believes in science," while Bush defended his decision to withdraw from Kyoto because "... it would have cost America a lot of jobs." The exchange ended when Kerry conceded that "The fact is that the Kyoto treaty was flawed" and criticised Bush for not trying "to fix it. He just declared it dead, ladies and gentlemen, and we walked away from the work of 160 nations over 10 years." Kerry failed to tell the audience how he proposed to "fix" Kyoto.

Public apathy partly explains the lack of attention given to climate change in the 2004 election. Although the movie *The Day After Tomorrow* and Michael Crichton's novel *State of Fear* popularised the issue, opinion polls revealed that environmental issues did not rank highly in lists of the public's concerns. War, terrorism, and the affordability of health care troubled Americans far more than climate change in 2004. Senator Kerry's failure to attack President Bush more aggressively on climate change also reflected, however, Republican success in framing the issue to their advantage. Bush's consistent characterisation of Kyoto as "flawed" and his claim that climate change was a long-term problem put the Democrats on the defensive and forced them to discuss the issue on their opponents' terms. Kerry not only struggled to make a case that the problem was immediate but also failed to counter his opponent's claims about economic costs and effectiveness. His efforts to portray the Bush Administration as anti-science had little impact in the short-term. Bush easily sidestepped such criticisms by acknowledging that climate change was a problem, and ensuring that the Party Platform contained no overt references to scientific scepticism. Subsequent revelations about the extent of the administration's interference in the scientific process would make it easier for Democratic candidates to attack the Republicans on this point in the 2008 elections.

The Second Term

President Bush's re-election in November 2004 meant little immediate change in climate change policy. During a visit to Europe in February 2005 Bush reiterated his position on the issue in a number of speeches and press conferences. In remarks made in Brussels, Belgium, on 21 February 2005, for example, he confirmed that the United States was committed to "addressing the serious, long-term challenge of global climate change" and argued that "emerging technologies ... will encourage economic growth that is environmentally responsible." Although no major departure from this basic position occurred over the next four years, the devastating impact of Hurricane Katrina, growing concerns about America's energy security, increased scientific evidence about climate change, and Democratic majorities in Congress following the 2006 midterm elections, prompted a number of changes at the margins. First, images of the devastation in New Orleans caused by Hurricane Katrina in August 2005 prompted significant media speculation about whether such storms were a consequence of climate change and generated increased attention to the issue. Second, efforts to address the country's dependence on imported oil offered new ways to frame the issue, and had "spillover" policy effects (Kingdon, 2011). Third, the growing weight of scientific evidence about climate change forced Bush to acknowledge the issue in his 2007 and 2008 State of the Union Addresses and consequently gave public prominence to the issue. Fourth, Democratic control of Congress provided opponents of the administration's policy on climate change with increased opportunities to attack Bush's record and offer proposals of their own. These changes in the "problem" and "political" streams were not substantial enough to open a "policy window" that would allow a major change in direction while President Bush remained in office, but were sufficient to produce minor changes in policy that provided the basis for further action under President Obama.

Media coverage of the aftermath of Hurricane Katrina showed a flooded landscape that looked as if it belonged in a science fiction movie. Images of the submerged streets of New Orleans and "environmental refugees" fleeing the disaster appeared to confirm what some climate scientists had been predicting for decades (Weart, 2008, 185; Booker, 2010, 136–7). Even before the hurricane struck New Orleans stories began appearing in the media positing a connection with climate change. A *Time* headline on 29 August 2005 asked "Is global warming fueling Katrina?" The next day a story in *The Boston Globe* answered the question when it declared that: "The hurricane that struck Louisiana yesterday was nicknamed Katrina by the National Weather Service. Its real name is global warming" (Gelbspan, 2005). Climate scientists were less certain that climate change had increased the frequency of hurricanes and bitter disputes raged about the possibility of any linkage (Pielke, 2010, 2014). The niceties of such disputes, however, meant little to a public fed images of an apparent *climatic* disaster. Hurricane Katrina provided a focus for public fears about climate change that proponents of action quickly seized. Images of post-Katrina New Orleans appeared in Al Gore's 2006 documentary *An Inconvenient Truth*, and posters advertising the film showed a cyclone emerging from the chimneys of a power station. Support for action on climate change began to grow as Americans started to believe that the consequences of inaction might affect Americans in the short-term rather than foreigners in the long-term. Gallup polls show that the percentage of Americans who believed that climate change was occurring rose from 12 per cent in 2004 to 25 per cent in 2007 (Dunlap, 2008).

The Bush Administration eventually responded to this change in the public mood with a number of measures ostensibly designed to improve the country's energy security. President Bush flagged energy security as an issue in his State of the Union Address given on 2 February 2005 when he urged Congress to enact the energy proposals that he had made early in his first administration to make "America more secure and less dependent on foreign energy." Using familiar language he stressed the need for "reliable" and "affordable" energy, but in a departure from his earlier rhetoric he also mentioned the need for "environmentally responsible energy." Bush returned to the subject of energy security in his State of the Union Address given on 31 January 2006. He argued that: "Keeping America competitive requires affordable energy. And here we have a problem. America is addicted to oil, which is often imported from unstable parts of the world. The best way to break this addiction is through technology." He called for increased spending on research into clean energy that would "change how we power our homes and offices" and "how we power our automobiles." The stress on technology as a solution was familiar but Bush proceeded to make remarks that belied his background as a Texas oilman. He claimed that "By applying the talent and technology of America, this country can dramatically improve our environment, move beyond a petroleum-based economy, and make our dependence on Middle Eastern oil a thing of the past."

In his 2005 and 2006 State of the Union Addresses President Bush had linked energy security and environmental improvement without mentioning climate change. Publication of the Stern Review on the Economics of Climate Change in October 2006 and knowledge that the 4th IPCC Assessment Report on climate change would be published in March 2007 made it difficult for Bush to continue avoiding the issue in his most important annual set piece speech. In his State of the Union Address given on 23 January 2007 he finally mentioned climate change. "America is on the verge of technological breakthroughs that will enable us to live our lives less dependent on oil. And these technologies will help us be better stewards of the environment, and they will help us to confront the serious challenge of climate change." Although Bush did not suggest that action should be taken to tackle climate change, and clearly focused on the need to enhance the country's energy security, the reference to the issue in the State of the Union Address did signal a modest shift in policy. A day after the State of the Union Address Bush issued Executive Order 13423 which required federal agencies "to improve energy efficiency and reduce greenhouse gas emissions." EO13423 set goals for reductions in energy consumption and the use of renewable fuels. Four months later Bush issued Executive Order 13432 which declared that "the policy of the United States" is "to protect the environment with respect to greenhouse gas emissions from motor vehicles," albeit "in a manner consistent with sound science, analysis of benefits and costs, public safety, and economic cost." Finally, Bush signed the Energy Independence and Security Act (2007) on 19 December 2007 which required automobile manufacturers to improve the fuel economy of cars by 40 per cent by 2020 and promoted a number of energy efficiency initiatives. In his signing statement Bush claimed that "The legislation I'm signing today will lead to some of the largest CO_2 emission cuts in our Nation's history" and will "make a major step toward ... confronting global climate change."

President Bush's mention of climate change in the 2007 State of the Union Address, together with his executive orders which recognised the problem and support for legislation that required new fuel economy standards to be set, represented tentative first steps towards a change of policy. Little sign of change, however, was apparent in other areas. The administration not only continued both to urge Congress to open access to domestic energy sources and to reject calls for an international treaty containing mandatory reductions in greenhouse gas emissions, but also opposed efforts by the states to regulate carbon dioxide. Two initiatives, in particular, posed a challenge to the administration's position on climate change. The first involved a legal challenge by 12 states and several cities to a 2003 EPA decision that greenhouse gases were not air pollutants as defined by the Clean Air Act (1990) (Martel and Stelcen, 2007,138). The case eventually reached the US Supreme Court which ruled in *Massachusetts* v. *EPA* (2007) that the EPA did have the authority to regulate greenhouse gases under the Clean Air Act, and needed to review its earlier decision not to regulate. The Court stated that the EPA had the discretion not to regulate, but such a decision needed to be grounded in the Clean Air Act rather than political considerations (Engel, 2010; Carlane, 2010). Reluctant to start a process that might lead to the regulation of greenhouse gases the Bush Administration decided to ignore the Court's decision and "run out the clock" (Rabe, 2010,16; Daniels, 2011). The second challenge flowed from a law passed in California in 2002 requiring automobile manufacturers to reduce emissions of greenhouse gases. For the law to take effect the Clean Air Act required the EPA to grant California a waiver allowing the state to have stricter air pollution control standards than mandated by the federal government. California requested a waiver in December 2005, but after considerable delay the EPA eventually denied the request. In December 2007 EPA Administrator Stephen Johnson argued that California was not uniquely affected by global warming, and therefore lacked the "compelling and extraordinary" conditions required under the Clean Air Act for a waiver to be granted (Carlane, 2010). California and a number of other states that wished to adopt the same standards promptly challenged the EPA's decision in the courts, and resolved to make another petition for a waiver after the 2008 elections.

The Bush Administration continued to reject calls for a new international treaty mandating limits on greenhouse gas emissions. Notice of "business as usual" in the international arena came in September 2007 when the administration organised a summit in Washington, DC, to explore alternative approaches to dealing with climate change (Rahm, 2010). In his remarks to delegates on 28 September 2007 Bush argued that: "We must lead the world to produce fewer greenhouse gas emissions, and we must do it in a way that does not undermine economic growth or prevent nations from delivering greater prosperity for their people." He called for a flexible international agreement with no mandatory reductions in greenhouse gas emissions. Demands that any successor to the Kyoto Protocol must avoid mandatory targets for reductions in greenhouse gas emissions, not harm economic growth, and apply to developed and developing countries alike, formed the basis of the American negotiating position at the 13th COP held in Bali, Indonesia, in December 2007 to decide a "roadmap" for negotiating a new climate change treaty. Frustration with these demands led delegates to boo the American negotiators at times but eventually a compromise was agreed (Jamieson, 2014, 52; Baer, 2012, 132). The Bali Action Plan called for "deep cuts" in emissions rather than detailing specific targets, and established a two-track process for negotiating a new treaty with countries wishing to avoid mandatary targets in one track and those wishing to build on the Kyoto Protocol in the other. In his State of the Union Address given on 28 January 2008 President Bush gave an indication of the type of treaty he favoured. He called for the creation of a "new international technology fund, which will help developing nations like India and China make a greater use of clean energy sources" and gave a warning that any "agreement will only be effective if it includes commitments by every major economy and gives none of them a free ride." Countries that favoured building upon Kyoto hoped that the two-track process agreed at Bali would enable negotiations to continue in the hope that the 2008 presidential elections would produce a new president with different views on climate change (Fuller and Revkin, 2007).

President Bush made his last major speech on climate change on 16 April 2008. This speech revealed some movement in his attitude towards the issue since the early years of his administration, but also reiterated familiar themes about economic growth, technology, mandatory standards, and international cooperation. Most notable is that Bush abandoned the overt scientific scepticism that had dominated his initial approach to the problem and acknowledged that action needed to be taken to address climate change. He claimed that "Climate change involves complicated science and generates vigorous debate. Many are concerned about the effect of climate change on our environment. Many are concerned about the effect of climate change policies on our economy. I share these concerns and believe they can be sensibly reconciled." He announced a new national goal to stop the growth of US greenhouse gas emissions by 2025 to be achieved by stricter fuel economy standards, improved energy efficiency, and the development of clean energy technology. Faith that technology would solve the problem remained a cornerstone of his approach to climate change and he denounced alternative ways of addressing the problem. "The wrong way is to raise taxes, duplicate mandates, or demand sudden and drastic emission cuts that have no chance of being realized and every chance of harming our economy," he warned. He also confirmed his opposition to any international agreement that allowed developing countries to "free ride." Bush insisted that he offered "a rational, balanced approach" to climate change. Pressure for a fundamental change in approach, however, had been growing for a number of years in both Congress and a number of states. Legislative initiatives to mandate reductions in greenhouse gas emissions had been introduced in Congress and several states had taken innovative action to tackle the problem. The administration had begun to lose control of the policymaking arena as proponents of change sought out alternative venues to take action.

Legislative Stirrings

Indicators of congressional activity show an increasing engagement with climate change during the Bush Administration as legislators reacted to growing scientific evidence about the problem, rising public concern, and presidential decisions. The number and range of bills addressing climate change rose dramatically during the period, committees conducted more hearings to examine the issue and discuss legislation, and a small number of measures reached the floor of the Senate for debate and a vote. The partisan composition of Congress, however, frustrated efforts to pass legislation and limited the scope of investigations into the actions of the administration for much of the period. Republican control of the House of Representatives from 2000 to 2006, and the Senate from 2002 to 2006, gave climate change sceptics institutional power as well as the numbers to defeat legislative proposals. Rep. W.J. "Billy" Tauzin (R. LA) chaired the House Committee on Energy and Commerce from 2000 to 2004 when he was replaced by the climate change sceptic Rep. Joe Barton (R. TX) who served as chair until 2006, and Senator James Inhofe (R. OK), a prominent climate change sceptic, chaired the Senate's Committee on Environment and Public Works from 2002 to 2006. The stranglehold held by climate change sceptics began to weaken when the Democrats gained majorities in both the House of Representatives and the Senate following the 2006 midterm elections. Although the slender size of the Democratic majority in the Senate (51–49) made it impossible to enact climate change legislation, majority status in both chambers provided Democrats with opportunities to challenge the administration's handling of the issue. Speaker Nancy Pelosi (D. CA) created a Select Committee on Energy Independence and Global Warming, chaired by Rep. Ed Markey (D. MA), to help shape policy, and control of the Senate's Committee on Environment and Public Works passed to Senator Barbara Boxer (D. CA) who had a strong record of supporting environmental causes. Institutional roadblocks to action remained, however, in the shape of Rep. John Dingell (D.MI), chair of the House Energy and Commerce Committee, and Senator Robert Byrd (D. WV), Chair of the Senate Appropriations Committee. Dingell had long championed the cause of the automobile industry while Byrd had been a co-sponsor of the Byrd-Hagel Resolution of 1997 and remained a strong defender of the coal industry.

President Bush's letter of 13 March 2001 to four Republican senators stating that he opposed the regulation of carbon dioxide emissions from power stations and opposed the Kyoto Protocol was greeted with dismay by many Democrats and some moderate Republicans (Lee et al., 2001). Opponents made statements denouncing the administration's position and introduced legislation offering alternative ways forward. Senator Kerry (D. MA) made a speech on the floor of the Senate on 14 March 2001, for example, in which he expressed concern about "the rather stunning announcement on the front page of a number of newspapers about President Bush's reversal of a campaign promise he made with great clarity in the course of the last year" (*Congressional Record*, S2300). Bills requiring reductions of carbon dioxide emissions (and other pollutants) from power stations were introduced by Senators James M. Jefford (R. VT), Patrick J. Leahy (D. VT), and Thomas R. Carper (D. DE), and Representatives Henry A. Waxman (D. CA), Thomas H. Allen (D. DE), and Charles H. Taylor (R. NC). Senator Jefford's bill was eventually reported favourably by the Senate's Committee on Environment and Public Works in June 2002 but no subsequent action was taken. No action was taken on the other bills. The most significant challenge to the administration's repudiation of Kyoto came in the form of an amendment offered by Rep. Robert Menendez (D. NJ) to the Foreign Relations Authorization Act for FY 2002–2003 that expressed the "Sense of Congress" that the United States should continue to participate in international negotiations to complete the rules and guidelines for the Kyoto Protocol. The House Committee on International Relations accepted the amendment on a 23–20 vote (five Republicans were absent) and the bill passed the House by a vote of 352–73 with no challenges to the Menendez Amendment. The amendment was subsequently removed from the bill, however, during conference discussions with the Senate at the insistence of the House Republican leadership.

Although the congressional challenge to President Bush over the regulation of carbon dioxide emissions and Kyoto ultimately petered out, the support given to legislative initiatives on climate change in the 107th Congress (2001–2002) revealed a degree of bipartisanship that indicated general support for action to further research, improve measurement of greenhouse gas emissions, and promote methods of carbon sequestration. An amendment to the Budget Resolution for FY2002 offered by Senator Kerry that added $4.5 billion to the federal budget to fund a range of climate change measures, for example, passed the Senate by voice vote in April 2001 with no opposition recorded. Support for a comprehensive approach to deal with climate change, however, did not exist. Many Republicans remained sceptical about the need to address the issue and wary of challenging President Bush while divisions between Democrats representing energy-producing states and those from energy-consuming states made consensus-building difficult within the party. Inter-party and intra-party divisions, together with split control of Congress, ruled out comprehensive action on climate change in the 107th Congress despite some signs of growing engagement with the issue.

Republican victories in the midterm elections of November 2002 ended prospects of a significant challenge to the Bush Administration's policy on climate change. Majority status in both the House and the Senate in the 108th Congress (2003–2004) strengthened the position of Republican opponents of action to tackle climate change. Republicans in the House proved unwilling to back proposals that had garnered a degree of bipartisan support in the 107th Congress. The House Committee on Energy and Commerce rejected an amendment offered by Rep. Henry Waxman (D. CA) to an energy policy bill (H.R. 6) that urged the United States to participate in international negotiations on climate change, and removed a similar amendment by Rep. Menendez from the Foreign Relations Authorization Act for FY2004. The Committee also voted to reject another Waxman amendment to H.R. 6 that required the president to establish a voluntary programme to reduce the carbon intensity of the United States by 18 per cent by 2012. Rejection of the latter amendment showed the extent of Republican opposition in the House to action on climate change as it proposed the same carbon intensity target that President Bush had announced on 14 February 2002 and granted no new regulatory authority.

Republican opposition to action on climate change was not as pronounced in the Senate as in the House. Although the appointment of Senator Inhofe as Chair of the Senate's Environment and Public Works Committee constituted a significant obstacle to action, those seeking to address climate change found other institutional venues to process policy and a degree of bipartisan support for selected initiatives on climate change. Senator Inhofe made his feelings about climate change clear on 28 July 2003 when he made a speech in the Senate in which he claimed that global warming is "the greatest hoax ever perpetrated on the American people" (*Congressional Record*, S10022), and his position as Chair of the Senate's prime committee for dealing with environmental issues gave him considerable authority to control the agenda. The appointment of Senator John McCain (R. AZ) as chair of the Senate Committee on Commerce, Science and Technology, however, offered those advocating action to tackle climate change an alternative venue to advance their agenda. McCain had developed a keen interest in climate change following the 2000 presidential primary elections and had forged close ties with Senator Joe Lieberman (D. CN) on the issue (Samuelsohn, 2009). On 3 August 2001 both Senators had made speeches in the Senate calling for action to limit greenhouse gas emissions, and in January 2003 they joined forces to introduce legislation (The Climate Stewardship Act of 2003) which proposed a cap-and-trade system for greenhouse gas emissions. When Senator Inhofe failed to hold hearings on the bill, McCain organised Commerce Committee hearings to explore the issue and negotiated a deal with Majority Leader Bill Frist (R. TN) to allow a floor debate and vote on the legislation.

The debate on The Climate Stewardship Act of 2003 on 29–30 October 2003 revealed familiar arguments over climate science and economic costs. Supporters of the legislation concentrated on presenting the scientific case for climate change while opponents raised doubts about the scientific evidence and claimed that action to limit greenhouse gas emissions would harm the American economy

(Besel, 2007). "I can hear the giant sucking sound of jobs leaving our country every time I return to Ohio" stated Senator George V. Voinovich (R. OH) during the debate on 30 October 2003 (*Congressional Record*, S13582). President Bush made his position clear when the White House issued a "Statement of Administration Policy" on 29 October 2003 in which he likened the proposed bill to the Kyoto Protocol, claimed that it would raise energy costs and harm manufacturing interests, and was inconsistent with the administration's strategy on climate change. In the floor vote on 30 October 2003 the Senate rejected the bill by a vote of 43–55 with 37 Democrats and 6 Republicans voting yes, and 46 Republicans and 9 Democrats voting no. Senator Inhofe interpreted the vote as showing that: "The science underlying this bill has been repudiated, the economic costs are too high, and the environmental benefits are non-existent" (Jalonick, 2003). Many environmentalists, on the other hand, expressed surprise that the bill secured the support of 43 senators and calculated that a further 5 or 6 senators might have voted for the measure but for last minute lobbying by the utility, automobile and mining industries (Little, 2003). A letter from a number of powerful trade associations urging senators to vote against the Climate Stewardship Act was sent just before the vote (Layzer, 2007).

Efforts to build upon the support for comprehensive action on climate change shown in the Senate vote on The Climate Stewardship Act of 2003 followed in the 109th Congress (2005–2006). Senators McCain and Lieberman introduced a revised version of their proposal (The Climate Stewardship Act of 2005), and Senators James Jeffords (I. VT), John Kerry (D. MA) and Jeff Bingaman (D. NM) also introduced bills to cap greenhouse gas emissions and establish a market in tradable permits. Continued Republican control of key committees, however, ensured that no action was taken on any of these initiatives. McCain again tried to circumvent the blocking power held by Senator Inhofe as Chair of the Environment and Public Works Committee by offering a version of The Climate Stewardship Act of 2005 as an amendment during floor debate on the Energy Policy Act of 2005 on 22 June 2005. The amendment was defeated by a vote of 38–60. Republican gains in the 2004 elections meant less support for the measure than in the previous Congress. Nonetheless a degree of bipartisan support for symbolic action on climate change remained. Just after rejecting McCain's amendment the Senate agreed to an amendment offered by Senator Bingham expressing the "Sense of the Senate" that Congress should enact a cap-and-trade programme to reduce greenhouse gas emissions as long as this did not harm the economy. Eleven Republicans voted against a motion to table (reject) the measure before it was passed by voice vote.

Republican antipathy towards climate change legislation continued in the House of Representatives during the 109th Congress. Representatives Wayne Gilchrest (R. MD), Tom Udall (D. NM), Henry Waxman (D. CA), and Jay Inslee (D. WA) introduced legislation to cap greenhouse gas emissions and establish a market for permits but no action was taken on any of the measures. Rep. Mike Fitzpatrick (R. PA) also introduced a "Sense of the House of Representatives" resolution similar to that which had passed the Senate but obtained no support. Parliamentary rules which restricted floor amendments during debate further limited opportunities to bypass the stranglehold that party leaders and committee chairs had over the policy process. Rep. Gilchrest was denied the opportunity to offer an amendment to the Energy Policy Act of 2005 that would establish a national greenhouse gas registry, for example, by the House Rules Committee which determines the conditions for floor action in the House. Opponents of action on climate change also used parliamentary rules to remove an amendment to an appropriations bill that expressed the sense of Congress that legislation mandating market-based limits on greenhouse gas emissions should be enacted. The amendment offered by Rep. Norman Dicks (D. WA) had been accepted by the House Appropriations Committee but was struck during floor debate on a point of order that it was not germane to the legislation under consideration.

Democratic victories in the 2006 midterm elections gave the party control of both the House and Senate for the first time since 1994 and changed the political landscape surrounding climate change. Supporters of action on climate change assumed a number of important institutional positions that enhanced their ability to control the agenda and use parliamentary rules to their advantage. Rep. Nancy Pelosi (D. CA) was elected speaker; Rep. Henry Waxman (D. CA) became chair of the House Committee on Oversight and Government; and Senator Barbara Boxer (D. CA) became chair of the Senate Committee on Environment and Public Works. The fact that President Bush remained in office, the small size of the Democratic majority in the Senate, and the institutional power of opponents of action such as Rep. John Dingell and Senator Robert Byrd meant that passage of comprehensive climate change legislation was impossible, but opportunities existed to shape the agenda, challenge presidential actions, and develop policy initiatives. The number of bills introduced in the 110th Congress that dealt with climate change more than doubled compared with the 109th Congress, congressional investigations into the administration's treatment of climate scientists began, and important policy changes concerning climate change were enacted in the Energy Independence and Security Act of 2007.

Speaker Pelosi signalled her intention to make climate change a priority in her early days in office when she created a new Select Committee on Energy Independence and Global Warming chaired by Rep. Ed Markey (D. MA) (Peters and Rosenthal, 2010, 72–3). The move was opposed by Rep. John Dingell who feared that the new committee would threaten the jurisdiction of the House Energy and Commerce Committee but Pelosi countered by insisting that Markey's committee would focus on exploring issues rather than processing legislation. Pelosi also introduced a major energy bill (New Direction for Energy Independence, National Security, and Consumer Protection Act) that included numerous provisions pertaining to climate change. The bill passed the House in April 2007 and many of its provisions on climate change were incorporated in the Energy Independence and Security Act of 2007 signed into law by President Bush on 19 December 2007. Other champions of action to address climate change also used their new institutional power in the House to push their agenda. Rep. Waxman pushed legislation through the House Committee on Oversight and Government that required federal agencies to reduce greenhouse gas emissions and meet new energy efficiency standards. Many of these requirements were also later incorporated in the Energy Independence and Security Act of 2007. Another prominent example involved Rep. Norman Dicks, chair of the House Appropriations Committee's Subcommittee on Interior, Environment, and Related Agencies Appropriations, who added a provision to the appropriations bill going through his subcommittee declaring that "it is the sense of Congress that there should be enacted a comprehensive and effective national program of mandatory, market-based limits and incentives on emissions of greenhouse gases." In the 109th Congress a similar provision had been removed by the Republican majority on a point of order during floor debate. The Democratic majority in the 110th Congress, however, defeated an attempt by Rep. Joe Barton (R. TX) to remove the amendment during floor debate and the measure passed the House.

The fragile Democratic majority in the Senate in the 110th Congress limited opportunities for major action to address climate change despite the presence of climate change advocates in important institutional positions. Supporters of action simply did not have the necessary majority to stand a chance of invoking cloture and winning a floor vote even if a handful of Republicans could be persuaded to support legislation as defections by Democratic senators representing industrial and energy producing states would negate such gains. The fate of The Lieberman-Warner Climate Security Act of 2007 bears witness to this political reality. Senators Joe Lieberman (I. CN) and John Warner (R. VA) introduced legislation to establish a cap-and-trade system for reducing greenhouse gas emissions in October 2007, and the bill was referred to the Senate Environment and Public Works Committee's Subcommittee on the Private Sector and Consumer Solutions to Global Warming and Wildlife Protection chaired by Senator Lieberman and where Senator Warner served as ranking minority member. Lieberman and Warner used their institutional position to report the bill and it passed to the full committee for consideration. The Environment and Public Works Committee finally voted 11–8 to report a version of the bill after a nine-hour hearing in May 2008 in which Senators Inhofe (R. OK) and Larry Craig (R. ID) tried to block progress by offering over 150 amendments. The reported version of the bill became known as The Boxer-Lieberman-Warner Climate Security Act of 2008 in recognition of the role played by Senator Barbara Boxer (D. CA), chair of the committee, in steering the bill through the committee. Senate Majority Leader Harry Reid (D. NV) agreed to schedule floor action but first had to overcome a filibuster organised by Senator Inhofe against the motion to proceed to debate. A cloture vote of 74–14 on 2 June 2008 allowed debate to begin. Opponents of the legislation made familiar claims about the economic cost of the proposals and employed a wide variety of obstructionist tactics to block the bill. These included requiring the Clerk of the Senate to spend more than 10 hours reading out the entire 491 page bill. On 6 June 2008 a cloture vote of 48–36 fell 12 votes short of the 60 votes needed to end debate and allow a vote on the measure. Seven Republicans voted for cloture and four Democrats voted against.

The record of the 110th Congress revealed both an increased willingness to address climate change but also continued obstacles to action. The number and range of measures introduced in both the House and the Senate showed high levels of interest in the issue and a growing propensity to deal with the issue from a variety of directions. Bills included comprehensive measures to limit greenhouse gas emissions by establishing cap-and-trade systems, resolutions calling for meaningful participation in international climate change negotiations, funding for the development of new technologies, and a range of energy efficiency measures. Provisions in the Energy Independence and Security Act of 2007 and other measures to promote energy efficiency, research into carbon capture and sequestration, and funding to help developing countries tackle the problem revealed a degree of consensus about the way forward based on a faith in technology. That consensus began to fray when proposals requiring changes in behaviour or imposing economic costs were under discussion. The number of bills proposing a cap-and-trade system, and the support garnered by the Lieberman-Warner Climate Security Act, however, provided some evidence of an emerging consensus about the advantages of reducing greenhouse gases in this way (Fisher et al., 2013). Defeat of Lieberman-Warner revealed continued obstacles to comprehensive action on climate change, nonetheless, particularly in the Senate where opportunities exist for a minority to block legislation. Numbers matter in legislative politics and the numbers were not there in 110th Congress to pass cap-and-trade legislation.

Conclusion

President Bush entered office in January 2001 committed to softening America's approach to dealing with climate change. In pursuit of this goal he repudiated the Kyoto Protocol and sought to stymie demands for further action by engaging in an aggressive effort to frame the issue in terms of scientific uncertainty, economic cost, and ineffectiveness. Changes in the problem, policy, and politics "streams" during his term in office, however, made this task somewhat akin to King Canute's efforts to hold back the tide. Scientific reports by international and domestic agencies made it difficult to maintain an argument that stressed scientific uncertainty, cap-and-trade appeared to offer a market-based solution to reducing greenhouse gas emissions, and the public mood seemed more willing to accept the need for government action. President Bush responded with a number of initiatives that promised improved energy efficiency and cleaner energy in keeping with the way that his two predecessors had sought to address the problem. Although Bush continued to resist pressure to set binding targets for greenhouse gas emissions, preferring to rely on voluntary action, other actors within the political system increasingly viewed such reductions as essential. Bills were introduced in Congress to establish a cap-and-trade system for reducing greenhouse gas emissions, which though not enacted, served to "soften up" elite and public opinion about the value of such an approach. By the time President Bush left office a consensus appeared to have formed that climate change posed a serious problem that needed to be addressed and that cap-and-trade was the best way to do this.

Chapter 5
Action and Reaction

Barack Obama's victory in the November 2008 presidential elections appeared to mark the beginning of a new era in climate change politics in the United States (Vidal, 2008). Obama had repeatedly stressed during the election campaign that climate change was a serious problem, argued that a cap-and-trade system was the best approach to reducing greenhouse gas emissions, promised leadership on the world stage, and seemed to have received a mandate for change. The Democrats had increased their majorities in Congress, advocates of cap-and-trade legislation had taken over the helm of key congressional committees, and public support for government action to address climate change appeared high. The promise of a new era in climate change politics quickly disappeared, however, as the old politics of scepticism, neglect, and obstruction re-emerged. Democratic support for climate change legislation in the Senate haemorrhaged away in the run-up to the 2010 midterm elections, Republicans regained control of the House of Representatives following those elections and launched assaults on regulatory initiatives taken in the early days of the administration, the exigencies of domestic and international politics hampered President Obama's efforts to play a leading role in treaty negotiations, and public support for action dissipated. Even Obama began to downplay his commitment to action on climate change in the face of economic and fiscal pressures, and the need to campaign for re-election in 2012.

The rapid re-emergence of the old politics of climate change during Obama's first administration meant that his main achievements occurred early in his term of office. President Obama issued a number of executive orders and directives in 2009 that reversed some of the decisions of his predecessor and opened the way for significant changes in the regulatory regime controlling greenhouse gases. These early achievements were marred, however, by the failure to enact legislation mandating reductions in greenhouse gas emissions. Legislation requiring cuts in greenhouse gas emissions passed the House of Representatives in June 2009, but died when the Senate failed to take action on its version of the bill before the midterm elections of 2010. The results of the November 2010 elections ended any prospect of climate change legislation, and allowed opponents of regulatory action to challenge Obama's early achievements. Congressional Republicans launched a number of initiatives to remove the EPA's authority to regulate greenhouse gas emissions and cut-off the funding for specific programmes related to climate change. The fiscal crisis provided further opportunities for broad attacks on environmental programmes in general and climate change action in particular. Four years after taking office President Obama found himself defending early achievements rather than taking new initiatives.

President Obama's re-election in 2012 offered him another opportunity to address climate change which he seized. In his Second Inaugural Address and subsequent State of the Union Address he highlighted the need to tackle climate change and promised to take executive action to reduce greenhouse gas emissions if Congress failed to act. His Climate Action Plan announced in June 2013 followed through on this promise. Obama revealed that he had directed the EPA to develop greenhouse gas emission standards for new and existing power stations that would significantly reduce emission levels. Congressional Republicans, and a number of Democrats from coal-mining states, reacted angrily to Obama's actions. They introduced legislation to remove the EPA's authority to take regulatory action and used the issue to their advantage in the 2014 midterm elections. The Republican capture of the Senate in these elections left Obama again defending his authority and added a further complication to his efforts to play a leading role in international climate negotiations.

The 2008 Election

Little attention was paid to climate change in the early stages of the 2008 presidential election nomination campaigns with most candidates focusing primarily upon foreign policy and domestic issues such as education, health, and the economy. References to climate change by Democrats tended to be brief while most Republicans ignored the subject completely. Hillary Clinton and Barack Obama occasionally acknowledged in their speeches that climate change was a serious threat and lambasted the Bush Administration for a lack of leadership on the issue but initially offered few concrete proposals of their own. Christopher Dodd talked vaguely of introducing a corporate carbon tax and Joe Biden, Dennis Kucinich, and Mike Gravel hardly mentioned the issue at all. Only Bill Richardson and John Edwards developed early proposals to address climate change and made the issue one of the central planks of their campaigns. Richardson constantly stressed the need to develop renewable energy sources and re-join the Kyoto Protocol in a number of statements in early 2007. Edwards announced an "Energy Plan" on 20 March 2007 which included a commitment to reduce greenhouse gas emissions by 80 per cent by 2050, introduce a cap-and-trade system for carbon dioxide, negotiate a new climate treaty, and support research and development of clean energy. Hillary Clinton and Barack Obama eventually responded with similar proposals of their own. In a press release on 20 April 2007 to announce that her campaign had gone "Carbon Neutral for Earth Day" Clinton also stated that she supported the use of a flexible, market-based system to combat global warming, but did not provide any details about what this meant. Obama announced that he supported a cap-and-trade system to reduce greenhouse gas emissions in a speech to the Chicago Council on Global Affairs on 23 April 2007. Among Republicans only John McCain displayed any engagement with the issue.

In a Speech on Energy Policy given on 23 April 2007, he stated that he viewed climate change as an important issue, supported a cap-and-trade system to reduce greenhouse gas emissions, and proposed collaborating with China to deal with the problem.

The emergence of cap-and-trade as the preferred tool for reducing emissions of greenhouse gases among leading Democratic candidates marked an important shift in thinking about how to deal with climate change. Previously Democrats had usually advocated a carbon tax as the best way to reduce energy consumption, and had regarded cap-and-trade as a Republican scheme designed to avoid the real costs of dealing with climate change (Broder, 2009a). To most Democrats a carbon tax had the advantage of directness and simplicity while cap-and-trade appeared counter-intuitive and complex. Democratic attitudes towards cap-and-trade began to change, however, as the success of the 1990 Clean Air Act Amendments cap-and-trade scheme for controlling sulphur dioxide became apparent, the Clinton Administration's 1993 proposal for an energy tax failed to secure congressional approval and contributed to the loss of the Democratic majority following the 1994 midterm elections, and perhaps most significantly, the dawning realisation that the initial distribution of permits could be used to garner political support. Most Republicans began to distance themselves from cap-and-trade, on the other hand, just as Democrats began to embrace the idea. Republicans began to attack cap-and-trade as something that would lead to bigger government, higher energy prices, and prove unworkable for carbon dioxide. John McCain was one of the few who continued to advocate the policy.

Democratic candidates developed detailed policy proposals and employed new rhetorical strategies to sell their ideas as the primary campaign progressed. John Edwards and Bill Richardson continued to stress the issue in their campaigns with the latter calling for an "American Revolution" to deal with the issue in a speech to the New America Foundation on 17 May 2007, and likening the effort to deal with climate change to the challenge of putting "men on the moon" in a press release on 3 June 2007. Hillary Clinton gave a commitment to reduce greenhouse gases by 80 per cent over the coming decade in a statement issued to mark Live Earth on 6 July 2007, and used rhetoric that mentioned the Apollo programme and the Manhattan Project in a speech discussing climate change in Portsmouth, New Hampshire, on 24 July 2007. Barack Obama developed his approach to climate change in a speech to the Detroit Economic Club on 7 May 2007 in which he stressed energy independence, becoming a leading player in the market for low carbon, technologies, and the green economy. Former Vice President Al Gore's receipt of the Nobel Peace Prize for his work highlighting the threat posed by global warming in October 2007 gave a further opportunity for Democratic candidates to emphasise their environmental credentials. Edwards and Richardson issued brief statements on 12 October 2007 congratulating Gore while highlighting their own records. Clinton gave a major speech on 5 November 2007 outlining a "Comprehensive Strategy to Address the Climate and Energy Challenge" in which she also praised Gore before discussing her own proposals and again describing climate change as "our space race." In the weeks leading up to the first caucuses and primaries in 2008 little divided the Democrats on climate change. All viewed it as an important issue, lambasted President Bush for his lack of leadership, used language that emphasised science, energy independence, the green economy, and national security, and agreed on the broad parameters of policy.

Republican candidates paid little attention to climate change as the campaign progressed with many having difficulty accepting that there was a problem. When pushed to address the issue most candidates equivocated or quickly changed the subject. The most telling example of this occurred during the debate between Republican candidates on 12 December 2007 in Johnston, Iowa. None of the candidates responded when the debate moderator asked them to raise their hands if they agreed that climate change was real. John McCain, Rudy Giuliani, and Mitt Romney eventually admitted that climate change might be real but Alan Keyes, and Fred Thompson refused to answer. Hillary Clinton exploited this episode in a debate between Democratic candidates three days later when she told the same moderator that she and her colleagues would be happy to raise their hands if asked the same question. Divisions between the Republican candidates began to emerge as the primaries and caucuses began. In the Republican debate held in Boca Raton, Florida on 24 January 2008 McCain stated that he believed that a cap-and-trade system would be the best way of reducing greenhouse gas emissions while Giuliani rejected the idea of any cap on emissions. Two days later Romney attacked McCain's position on climate change. In a press release titled "Leading the Charge on the Other Side," he claimed that McCain's policies were identical to those of the Democrats and would cost Americans $1,000 per year in higher carbon taxes. He returned to this theme in the Republican debate held in Simi Valley, California, on 30 January 2008 claiming that McCain's proposals would mean that Americans would pay all the cost of addressing global warming and noting that: "They don't call it American warming. They call it global warming." Ron Paul added his voice to the attacks on McCain at a speech to the Conservative Political Action Conference on 7 February 2008 when he asked: "our leading candidate, guess whose position he holds on global warming? Al Gore's. He supports the Al Gore bill on global warming."

The differences between the Democratic and Republican candidates in their approach to climate change during the contest for their respective party's presidential nomination reflected both partisan divisions among the electorate and intra-party divisions within the Republican Party. A Pew Research Center survey conducted in January 2007 revealed that roughly twice as many Democrats (48 per cent) as Republicans (23 per cent) viewed global warming as a "top priority" for government (Pew, 2007). These figures placed the issue near the bottom of the priorities list for Democrats, but right at the bottom for Republicans. Republican interest in climate change fell even further during the next 12 months with only 12 per cent regarding it as a priority issue in January 2008 (Pew, 2008). The two surveys also revealed considerable divisions among Republicans on almost every question related to global warming. Conservative

Republicans proved more likely to question the evidence that global temperatures were rising, to dispute claims that human activity was the cause of global warming, and to deny that the issue should be a "top priority" for government than moderate Republicans. While Democratic candidates could be fairly certain that addressing climate change in their campaigns for the party's nomination would not harm their prospects of success, Republican candidates had to tread far more carefully as the issue lacked a constituency within the party and had the potential to alienate potential supporters.

The discussion of climate change in the party platforms adopted at the start of the general election testify to the different calculations and political pressures that Barack Obama and John McCain faced when formulating policy positions on the issue. Although both Obama and McCain had stated in their nomination campaigns that climate change was real, needed addressing, and proposed similar cap-and-trade solutions, the party platforms differed considerably in their treatment of the issue. While the Democratic Party Platform adopted on 25 August 2008 stated unambiguously that, "Global climate change is the planet's greatest threat …," the Republican Party Platform adopted on 1 September 2008 stated that, "… the scope and longterm consequences of [the build-up of carbon in the atmosphere] are the subject of ongoing scientific research" and cautioned "… against the doomsday climate change scenarios peddled by the aficionados of centralized command-and-control government." The Democratic platform included commitments to implement a cap-and-trade system, invest in the green economy, and play a leading role in international efforts to combat global warming. The Republican platform contained commitments to "support technology-driven, market-based solutions" to reduce emissions, but noted that climate policies must "… not force Americans to sacrifice their way of life or trim their hopes and dreams for their children" or require "… the US to carry burdens which are more appropriately shared by all." Al Gore commented on the compromises that McCain needed to make to secure agreement on his party's platform in a speech at the Democratic National Convention in Denver on 28 August 2008 when he remarked that McCain "has apparently allowed his [party] to brow beat him into abandoning mandatory caps on global warming pollutants."

John McCain reasserted his commitment to tackling climate change during the general election as he sought to distance himself from the Bush Administration and persuade the public that he possessed the courage to make decisions that were politically unpopular. In the presidential debates held on 7 October 2008 and 15 October 2008 he stressed that he had disagreed with President Bush on the issue and had introduced legislation on the issue in the US Senate. Little separated McCain and Obama, in fact, on the issue as the campaign neared its end. Both agreed that climate change needed to be addressed, tended to frame the issue in terms of energy, and advocated a cap-and-trade system to reduce greenhouse gases. Tensions within the Republican Party on the issue, however, remained close to the surface. In the Vice Presidential Debate on 2 October 2008 in St Louis, Sarah Palin agreed that climate change was real but then questioned whether it was all due to human activity. Republican voters also revealed disquiet about McCain's position on climate change. One survey conducted in the second week of October 2008 revealed that 45 per cent of John McCain supporters distrusted him as a source of information about global warming (Leiserowitz and Mailbach, 2009). In contrast only 15 per cent of Obama supporters distrusted their candidate on the issue.

There is little evidence that climate change was a decisive issue in determining the outcome of the 2008 presidential election. Barack Obama defeated John McCain because his promise of "change" resonated with voters. The result of the election and those for the US Congress, however, produced sufficient changes in the "political stream" to suggest action on climate change was likely (Kingdon, 2011). First, the fact that Obama had highlighted both the issue and a means to deal with it during his campaign meant that his victory could be interpreted as a mandate for a change in policy. Environmental groups certainly believed this to be the case and greeted his election with euphoria and heightened expectations of action (Bomberg and Super, 2009). Second, the congressional elections returned increased Democratic majorities in the House of Representatives and the Senate. Democrats had a sufficiently large majority in the House to pass legislation even if some party members voted with the Republicans, and a filibuster-proof majority in the Senate assuming they could count on the votes of Senator Joe Lieberman (I. CN) and Senator Bernie Sanders (I. VT). Third, the interest group environment seemed more favourable with environmental groups refining their message on climate change and some large corporations advocating action to address the problem (Bryner, 2008). Finally, the public mood appeared to support action on climate change. Opinion polls conducted in 2007–2009 revealed high levels of public knowledge and concern about the problem (Dugan, 2014; Saad, 2014). In short, a new era in climate change politics appeared to have dawned.

Early Achievements

A confluence of political, problem, and policy "streams" opened a "policy window" for action on climate change in the period following the 2008 elections (Kingdon, 2011). Not only had the election altered the political stream, but the campaign had also offered different ways of framing climate change that linked the problem to traditional concerns such as energy independence, national security, and economic prosperity, and had additionally served to "soften-up" policymakers and the public about the merits of solutions such as cap-and-trade. Barack Obama sought to take advantage of this open window by moving rapidly during the transition period to advance his agenda on climate change. On 18 November 2008, less than two weeks after his electoral victory, he sent a taped video-message to a bipartisan group of state governors meeting in Los Angeles to discuss climate change that gave notice that: "My presidency will mark a new chapter in America's leadership on climate change that will strengthen our security and create millions of jobs," confirmed his support for both a cap-and-trade system to reduce greenhouse gas emissions and greater investment in green energy technologies, and concluded with the exhortation that: "Now is the time to confront this challenge once and for all. Delay is no longer an option. Denial is no longer an acceptable response. The stakes are too high." A month later Obama announced a number of appointments that signalled the seriousness of his desire to break with the previous administrations attitudes and policies towards climate change.

Barack Obama announced his key energy and environmental appointments on 15 December 2008. He appointed Steven Chu, a Nobel laureate at the University of California, Berkeley, as Secretary of Energy. Chu had a reputation as a campaigner on climate change as well as impeccable scientific credentials. Obama also appointed Carol Browner, a former administrator of the EPA during the Clinton Administration, to a newly created post of White House Co-ordinator for Energy and Climate Policy. Environmentalists praised both appointments. Gene Karpinski, head of the League of Conservation Voters, celebrated Chu and Browner as "a green dream team" (Goldenberg, 2008). Further significant appointments included Lisa Jackson, New Jersey's Commissioner of Environmental Protection as EPA Administrator, and Nancy Sutley, Deputy Mayor of Los Angeles for Energy and Environment, as head of the Council on Environmental Quality. In remarks announcing his appointments Obama claimed that: "The team I have assembled here today is uniquely suited to meet the great challenges of this defining moment. They are leading experts and accomplished managers, and they are ready to reform government and help transform our economy so that our people are more prosperous, our nation is more secure, and our planet is protected." Secretary of State Hillary Clinton's subsequent appointment of Todd Stern as the State Department's Special Envoy for Climate Change underscored the new administration's determination to break with the past. When announcing the appointment on 26 January 2009 Clinton stated that "we are sending an unequivocal message that the United States will be energetic, focused, strategic and serious about addressing global climate change and the corollary issue of clean energy" (Clinton, 2009).

President Obama emphasised the need for action to address climate change in his Inaugural Address on 20 January 2009 when he claimed that "the ways we use energy strengthens our adversaries and threaten our planet," referred to the "specter of a warming planet," and spoke of an ambition to "harness the sun and the winds and the soil to fuel our cars and run our factories." This rhetoric continued a framing strategy that had been evident during the nomination and general election campaigns when Obama and other Democrats had talked about energy independence, energy security, and green jobs in an effort to counter Republican charges that proposals to reduce greenhouse gas emissions would harm the American economy and damage the country's national interest. President Obama made a further effort to frame climate change along these lines in a major speech on 26 January 2009 that outlined a "New Energy for America" plan. Obama claimed that the "urgent dangers to our national and economic security" resulting from America's dependence on oil from overseas "are compounded by the long-term threat of climate change," and announced a number of policy initiatives to safeguard "our security, our economy and our planet." He promised new investment in alternative energy that would create 460,000 jobs, new fuel-efficiency standards for automobiles and trucks, a willingness to cooperate with the states to reduce greenhouse gas emissions, and American leadership to fashion "a truly global coalition" to protect the climate. Obama sought to sell his climate change policies with language that raised the prospect of hostile regimes controlling the country's energy supplies, promised new jobs in the midst of a recession, and downplayed the need for changes in consumption (Bomberg and Super, 2009). He talked about developing new sources of energy rather than reducing energy consumption, and driving more fuel-efficient automobiles rather than reducing automobile use.

President Obama issued two memoranda following his speech on 26 January 2009 that reversed decisions taken by his predecessor and paved the way for regulatory action to reduce greenhouse gas emissions. The first directed the Department of Transport (DOT) to finalise regulations requiring the automobile industry to improve fuel efficiency standards as mandated by the Energy Independence and Security Act of 2007 (Obama, 2009a), and the second required the EPA to reconsider California's request for a waiver to regulate greenhouse gas emissions from automobiles as allowed under the Clean Air Act of 1990 (Obama, 2009b). The prospect of having to meet different standards effecting fuel economy and emissions promulgated by the DOT and California, and the possibility that the EPA might consider further regulations restricting greenhouse gas emissions from automobiles under the Clean Air Act, led to urgent negotiations between the industry and federal and state officials. Struggling to survive as a result of the recession and dependent upon federal financial help, the automobile manufacturers gave way and accepted the need for tougher standards. On 19 May 2009 Obama announced that negotiations between the Department of Transport, the EPA, the automobile industry, and California had produced an agreement to tighten fuel efficiency standards and reduce greenhouse gas emissions from automobiles and light-duty vehicles (SUVs). "For the first time in history, we have set in motion a national policy aimed at both increasing gas mileage and decreasing greenhouse gas pollution for all new trucks and cars sold in the United States," Obama stated in his announcement, and articulated a goal "to set one national standard that will rapidly increase fuel efficiency … by an average of 5 per cent each year between 2012 and 2016." Environmentalists praised the new policy (Goldenberg, 2009b). Daniel Becker, director of the Safe Climate Campaign, claimed that: "This is the biggest step the American government has ever taken to cut greenhouse gas emissions" (Broder, 2009b). The EPA formally granted California's waiver to regulate greenhouse gas emissions from automobiles in June 2009, but as part of the agreement announced by President Obama the month previously, the state confirmed that any automobile manufacturer achieving new federal standards would also be deemed to be in compliance with state standards.

The legal basis for standards to regulate greenhouse gas emissions from motor vehicles flowed from the US Supreme Court's ruling in *Massachusetts* v. *EPA* (2007) that greenhouse gases met the definition of air pollutants under the Clean Air Act of 1990, and the EPA should determine whether greenhouse gases from new motor vehicles posed a public health risk (Daniels, 2011). In April 2009 the EPA issued a proposed Endangerment Finding that greenhouse gases contributed to climate change, and consequently posed a threat to the health and welfare of present and future generations. Nine months later the EPA confirmed this view. An Endangerment Finding issued on 7 December 2009 concluded that greenhouse gases from automobiles posed a threat to public health and welfare because they contributed to climate change. In a press release announcing the Finding, EPA Administrator Lisa Jackson took a swipe at the Bush Administration's failure to act upon the Court's ruling, and highlighted the importance of the Finding to President Obama's plans, when she commented that: "These long-overdue findings cement 2009's place in history as the year when the United States Government began addressing the challenge of greenhouse-gas pollution and seizing the opportunity of clean-energy reform," (USEPA, 2009). Several legal challenges followed the announcement of the Endangerment Finding but these have all failed. In June 2012 the US Court of Appeals for the DC Circuit ruled in *Coalition for Responsible Regulation* v. *EPA* that the EPA had the legal authority to issue the Finding.

The DOT and EPA proposed new harmonised fuel economy and emission standards for automobiles and light-duty vehicles in September 2009 and issued final regulations in April 2010. These regulations required reductions in greenhouse gas emissions from automobiles, equivalent to a 35.5 mpg fuel economy standard, to be achieved by 2016. The EPA and DOT claimed that these standards would reduce greenhouse gas emissions from automobiles and other light-duty vehicles by 21 per cent by 2030 (USEPA, 2010). Just over a month later President Obama issued a further memorandum directing the EPA and DOT to establish emission standards for medium and heavy trucks and also produce tighter standards for automobiles for the period 2017–2025 (Obama, 2010a). In a press release issued on 21 May 2010 Obama defended this call for new standards using a familiar framing strategy. He argued that: "This will bring down costs … it will reduce pollution … and will spur growth in the clean energy sector. We know how important this is. We know our dependence on foreign oil endangers our security and our economy. We know that climate change poses a threat to our way of life … And we know that our economic future depends on our leadership in the industries of the future." The DOT and EPA proposed new harmonised fuel economy and emission standards for medium and heavy trucks to cover the period 2014–2018 in October 2010 and issued final regulations in August 2011. The EPA and DOT estimated that the standards would reduce greenhouse gas emissions by 270 million metric tons over the life of vehicles built for the 2014–2018 model years (USEPA, 2011). Proposed new emission and fuel economy standards for automobiles and light-duty vehicles were announced three months later. These required reductions in greenhouse gas emissions from automobiles, equivalent to a 54.5 mpg fuel-economy standard, to be achieved by 2025.

The significance of the vehicle emission standards promulgated by the EPA and DOT lay not only in the reductions in greenhouse gas emissions they mandated, but also in the other regulatory action they triggered. Under the Clean Air Act of 1990 further regulation of a pollutant is required once that pollutant has been regulated under any part of the act. The fact that the legal authority for the harmonised emission and fuel economy standards issued in April 2010 came from the Clean Air Act, therefore, opened the way for broader regulation of greenhouse gases under this statute. Regulatory action followed quickly. In May 2010 the EPA determined that new or modified stationary sources of greenhouse gas emissions (power stations, factories) should be subject to the Prevention of Serious Deterioration (PSD) provisions of the Clean Air Act, and issued a "tailoring" rule that required the installation of "best available control technology" from January 2011 if such sources increased their emissions above a specific threshold. In December 2010 the EPA further agreed to promulgate New Source Performance Standards (NSPS) to regulate greenhouse gas emissions from power stations and refineries as part of a judicial settlement to end a number of lawsuits dating back to the Bush Administration. The EPA announced draft standards for power stations in March 2012 that would limit carbon dioxide emissions from new power plants to 1,000 pounds per megawatt-hour.

EPA Administrator Lisa P. Jackson claimed that the new standards marked a significant step in the battle to reduce greenhouse gas emissions: "Right now there are no limits to the amount of carbon pollution that future power plants will be able to put into our skies—and the health and economic threats of a changing climate continue to grow. We're putting in place a standard that relies on the use of clean, American made technology to tackle a challenge that we can't leave to our kids and grandkids" (USEPA, 2012). Reaction to the announcement, however, revealed familiar fault-lines. Supporters of the coal industry argued that the new standards would mean the end of coal-fired power stations as the technology to reduce emissions to the required levels did not exist commercially. Power stations fuelled by natural gas, on the other hand, would easily be able to meet the standards. Democrats and Republicans from coal mining states moved quickly to criticise the proposed rules. Senator Joe Manchin (D. WV) claimed that: "This EPA is fully engaging in a war on coal, even though this country will continue to rely on coal as an affordable, stable and abundant energy source for decades to come" (Barringer, 2012).

Senator Rick Santorum (R. PA) made a similar comment while campaigning for the Republican presidential nomination in a press release dated 27 March 2012: "President Obama's environmental agenda kills American jobs, creates higher energy prices and weakens our nation's security. America is the Saudi Arabia of coal, and we could create our own energy if the government would let us." Environmentalists generally welcomed the proposed new rules but expressed disappointment that they did not apply to existing coal-fired power stations. Frances Beinecke of the Natural Resources Defense Council stated that: "The logical next step is to improve the aging fleet of existing coal-fired power plants, which remain the major source of industrial carbon pollution in our country" (Barringer, 2012).

President Obama supplemented the rule-making triggered by the two presidential memoranda issued in January 2009 with other initiatives requiring action within the federal government (Wold, 2012; Adler, 2011). On 5 October 2009 he issued an important Executive Order on Federal Sustainability (EO 13514) that required the federal agencies to set greenhouse gas emissions reduction targets within 90 days. Noting that the federal government owned 600,000 vehicles, owned and managed 500,000 buildings, and paid $24.5 billion in utility and fuel bills in 2008, Obama argued that the government needed to reduce its fuel consumption and set an example that others could follow. On 29 January 2010 he announced that the federal government would reduce direct emissions from fuel and building energy use by 28 per cent by 2020. On 20 July 2010 he further announced that the federal government would reduce indirect emissions by 13 per cent. EO 13514 provided the authority for another presidential memorandum with implications for greenhouse gas emissions. On 24 May 2011 Obama issued a Memorandum on Federal Fleet Performance which established a target for the federal government to purchase only alternative fuel vehicles by 2015.

Environmentalists welcomed the various executive and regulatory actions taken by the Obama Administration to address climate change. Writing in the *Los Angeles Times* Michael Brune, the executive director of the Sierra Club, praised the administration's efforts to revive the EPA, toughen fuel-economy standards, and "most significantly, declaring that under the Clean Air Act, greenhouse gases endanger the public health and welfare" (Brune, 2011). Brune and others expressed disappointment, however, that these achievements had not been mirrored in the legislative arena (Walsh, 2011). Al Gore criticised Obama for his inability to drive legislation mandating reductions in greenhouse gas emissions through Congress, for example, in a prominent 7,000 word essay published in *Rolling Stone* magazine (Gore, 2011). Although the American Recovery and Reinvestment Act of 2009 (ARRA) fulfilled Obama's campaign promise to invest in the green economy, he proved unable to fulfil his other campaign promise to establish a cap-and-trade system to reduce emissions of greenhouse gases before the midterm elections of 2010. Republican gains in those elections finally ended any prospect of legislation to reduce greenhouse gas emissions during Obama's first term in office, and provided climate change sceptics with enhanced opportunities to attack executive initiatives. Bills seeking to strip the EPA of its authority to regulate greenhouse gases, and remove funding for personnel and programmes dealing with climate change, became commonplace and forced Obama onto the defensive.

Legislative Failure

The likelihood of the 111th Congress (2009–2011) enacting legislation to combat climate change appeared high in January 2009. Democrats had increased majorities in both the House of Representatives and the Senate following the 2008 elections, and legislators with strong environmental credentials had assumed important leadership positions in both chambers. The size of the Democratic majority in the House (255 to 179) meant that the party had sufficient votes to enact legislation mandating reductions in emissions of greenhouse gases even if some members voted with the opposition. Numbers in the Senate were tighter but if the 58 Democrats maintained unity they could block Republican attempts to filibuster legislation with the support of two independents. House Speaker Nancy Pelosi (D. Ca); Henry Waxman (D. CA), the new chair of the House Energy and Commerce Committee; and Ed Markey (D. MA), chair of the Select Committee on Energy Independence and Global Warming and also chair of the House Energy and Commerce Committee's Subcommittee on Energy and Environment had long supported action on climate change. Waxman's ousting of John Dingell (D. MI) as chair of the Energy and Commerce Committee, in particular, augured well for action on climate change. While Dingell had long been a defender of the automobile industry, Waxman had a reputation as a champion of environmental issues. One spokesman for the automobile industry even described Waxman as an "irrational environmental zealot" (Broder, 2008). Senate Majority Leader Harry Reid (D. NV) had a good environmental record and could be expected to support Obama's agenda, Barbara Boxer (D.CA), chair of the Senate Environment and Public Works Committee, had a reputation as an environmentalist, and the decision of Robert Byrd (D. WV) to give up the chair of the Senate Appropriations Committee removed a champion of the coal-mining industry and co-author of the Byrd-Hagel Resolution of 1997 from a position of power.

President Obama made two policy commitments on climate change during his election campaign that required congressional action. First, he had consistently stressed the need to invest in alternative energy sources both as a means of securing the country's "energy independence" and as a way of reducing greenhouse gas emissions. The Democratic Party Platform contained a promise to "invest in energy technologies, to build the clean energy economy and create millions of new, good 'Green Collar' American jobs." In various speeches during the campaign and after his election Obama talked about investing $150 billion over 10 years in alternative energy. Second, he advocated a cap-and-trade system to reduce greenhouse gas emissions. The Democratic Party Platform included a commitment to "implement a market-based cap-and-trade system to reduce greenhouse gas emissions by the amount scientists say is necessary to avoid catastrophic change." Obama stated during the campaign that he proposed to reduce greenhouse gas emissions to 80 per cent below 1990 levels by 2050 and would establish strong annual reduction targets to ensure the United States achieved this figure.

The American Recovery and Reinvestment Act (ARRA) signed into law by President Obama on 17 February 2009 made good on the promise to invest in the green economy. This $800 billion economic stimulus package included $42 billion for energy-related investments, $21 billion for energy-related tax incentives, $1.6 billion for Clean Renewable Energy Bonds, and $570 million for climate science research. Provisions included funding or tax credits for the development of renewable sources of energy (solar, wind, hydro), power transportation systems (the smart grid), geothermal technology, new fuel-cell technologies, improved batteries, electric vehicles, and promoting energy efficiency. In remarks made when signing the law, Obama claimed that "we are taking big steps down the road to energy independence, laying the groundwork for new green energy economies that can create countless well-paying jobs. It's an investment that will double the amount of renewable energy produced over the next 3 years." Primarily intended to boost the economy a number of commentators viewed the ARRA as a major piece of environmental reform (Chait, 2013; Grunwald, 2012).

President Obama sought to take advantage of the momentum created by the passage of ARRA with a call for Congress to enact a cap-and-trade law to reduce greenhouse gas emissions. In an Address Before a Joint Session of Congress on 24 February 2009 he tried to persuade legislators that passage of such a law flowed naturally from the steps taken to promote energy independence and create a green economy in ARRA. "But to truly transform our economy, to protect our security, and save the planet from the ravages of climate change, we need to make clean, renewable energy the profitable kind of energy," he stated, "So I ask this Congress to send me legislation that places a market-based cap on carbon pollution and drives the production of more renewable energy in America." Obama also called for $15 billion a year investment in alternative energy. Two days later Obama provided some details of his cap-and-trade plan when he submitted his Budget for FY 2010 to Congress. The Budget revealed targets for reducing greenhouse gas emissions by 14 per cent below 2005 levels by 2020, by 83 per cent below 2005 levels by 2050, and included an income stream of $646 billion from auctioning emission permits over the next 10 years.

A number of factors made it unlikely that Congress would act with the same urgency in enacting cap-and-trade legislation as had been seen with ARRA. First, the public did not regard dealing with climate change as a high priority. An opinion poll published by PEW in January 2009 revealed that the number of respondents identifying global warming as a top domestic priority for President Obama and Congress had fallen from 35 per cent in January 2008 to 30 per cent in January 2009 (PEW, 2009). In contrast the number identifying "strengthening the nation's economy" as a high priority increased from 75 per cent to 85 per cent over the same period. Obama's framing strategy of linking economic regeneration and climate change made sense given these figures, but depended upon overcoming long-held views that environmental protection came at a cost to the economy. Second, the distributive nature of ARRA reduced interest group opposition to the measure. Business groups campaigned to secure benefits from the federal trough rather than defeat the measure. In contrast, Obama's cap-and-trade proposals had regulatory and redistributive elements that provided benefits to some but imposed burdens on others. The conflict generated by the proposal promised to be greater than over ARRA given the existence of winners and losers, and posed considerable challenges for law-making. Third, ARRA had passed on near party line votes in both the House and the Senate as President Obama had proved able to persuade just 11 House Republicans and three Senate Republicans to support the measure.

Scepticism about climate change and big government made it unlikely that Obama could count upon much support from Republicans to pass cap-and-trade legislation, and meant that he would have to rely upon Democratic votes to succeed. The need for a super-majority of 60 votes in the Senate to overcome any filibusters required all Democratic senators and the two Independents to support the measure. Finally, President Obama had focused his attention on ARRA and made clear that an economic stimulus package was his main priority. Cap-and-trade legislation would have to compete with health reform for his attention.

President Obama left the task of drafting cap-and-trade legislation to Rep. Henry Waxman, chair of the House Energy and Commerce Committee, and Rep. Ed Markey, chair of the committee's Subcommittee on Energy and Environment. On 31 March 2009 they released a "discussion draft" of the American Clean Energy and Security Act (ACES Act) for discussion in the Energy and Commerce Committee. This first version of the ACES Act proposed introducing a cap-and-trade system that would reduce emissions of greenhouse gases by 3 per cent below 2005 levels in 2012, 20 per cent below that level in 2020, and 83 per cent below 2005 levels by 2050. No details were given about how emission allowances would be allocated, but the proposed legislation did include an offset scheme which allowed industries to offset emissions of greenhouse gases against other actions that reduced the amount of carbon dioxide in the atmosphere. The draft also included provisions that required utility companies to produce 6 per cent of their electricity from renewable sources by 2012 rising to 25 per cent by 2025, and boosted funding for research into carbon capture and sequestration, and the development of clean fuel and vehicles, smart grid and electricity transmission systems, and energy efficiency.

Introduction of the "discussion draft" sparked a political frenzy as legislators sought to protect economic interests in their constituencies and ensure that voters suffered as little financial hardship as possible, while industrial, business, and agricultural groups sought changes that would increase any benefits resulting from the proposed legislation and minimise the costs of compliance. Reports of lobbying activity revealed a 50 per cent increase in spending by the seven largest fossil fuel companies, for example, as they sought to influence the outcome of legislation (Goldenberg, 2009a). Key areas of contention included the level of the cap on greenhouse gas emissions, how allowances would be allocated, the proportion of electricity that had to be generated from renewable sources, the system and value of the offset programme, and how to ensure that consumers would not be faced with rapidly escalating energy prices. Other arguments raged around the scope and level of federal spending contained in the proposed legislation. Numerous industry groups demanded help from the government to make the transition to a green economy while legislators sought federal funding for pet projects in their constituencies that often had little connection to climate change. The horse trading involved in securing and maintaining a Democratic majority for the proposed legislation in the face of general Republican opposition continued until passage in the House of Representatives.

Evidence of the compromises need to keep Democrats on board can be seen in the version of the ACES Act introduced by Henry Waxman and Ed Markey on 15 May 2009. The revised version reduced the target for greenhouse gas emissions to 17 per cent below 2005 levels by 2020, and set a new target for the production of electricity from renewable sources of 20 per cent by 2020. Perhaps most significantly, the revised version proposed to give away 80 per cent of initial emission allowances rather than sell them at auction. Waxman and Markey made this concession in order to secure the backing of Democrats from coal-mining areas who feared an auction would disadvantage coal-burning power stations, and to allay fears that utility companies might pass on increased costs to consumers. Other concessions, compromises and gifts to win the votes of wavering legislators and the support of powerful industries increased the length of the Bill from 648 pages to over a thousand pages. President Obama accepted these deals as the price of passage. In his Weekly Address on 16 May 2009 he praised Waxman and members of the Energy and Commerce Committee for bringing "together stakeholders from all corners of the country and every sector of our economy to reach an historic agreement on comprehensive energy legislation." Republicans on the House Energy and Commerce Committee offered a number of amendments during the committee's consideration of the ACES Act that would have removed key provisions or unpicked the compromises negotiated by the Waxman and Markey but all failed. The House Energy and Commerce Committee voted 33–25 to approve the ACES Act on 21 May 2009. One Republican voted in favour while four Democrats voted against.

Horse-trading and deal-making continued until the last moment (Broder, 2009d). In the period leading up to floor debate on 26 June 2009 Henry Waxman and Edward Markey made further concessions to maintain support. These revisions established a reduced target for the production of electricity from renewable sources of 15 per cent by 2020 and raised the proportion of initial emission allowances permits to be given away to 85 per cent of the total. Important deals on provisions effecting farmers and forestry companies ensured the support of Democrats representing rural constituencies. The last minute nature of many of these deals meant that Waxman had to offer a last minute substitute amendment to the Bill that added 300 pages to its length and included numerous "placeholders" where the details would be filled in later. This tactic incensed Republicans who claimed they had not been able to obtain, let alone read, the latest version of the Bill before being asked to debate and vote on the measure. Procedural objections came to nothing, however, and floor debate followed a familiar pattern. Democrats talked about energy independence, the green economy, and the need to address a potential environmental catastrophe while Republicans talked about the harm the measure would cause to the economy, the cost to consumers, and the prospect of a burgeoning federal bureaucracy that would interfere in everyday lives. The ACES Act eventually passed the House of Representatives on a 219 to 212 vote with 211 Democrats and 8 Republicans voting yes and 168 Republicans and 44 Democrats voting no.

President Obama lobbied hard both privately and publically in the weeks leading up to the floor vote on the ACES Act to secure its passage. Early in June 2009 he invited key members of the House of Representatives to the White House where he made an appeal for action on the bill that secured the support of wavering Democrats (Goldenberg, 2009c). This meeting was followed by others between administration officials and legislators in an effort to fashion a majority for the bill (Goldenberg, 2009d). In the week before the scheduled vote Obama supplemented this "insider strategy" with a "going public" strategy that sought to mobilise public support for the measure. At a news conference on 23 June 2009 he stressed themes of energy security, green jobs, and environmental catastrophe, and urged the House of Representative to pass the law with a clear statement that "this legislation is extremely important for our country." Two days later he stood in the Rose Garden and made an impassioned plea for passage that addressed criticism of the bill. In these Remarks on Energy Legislation delivered on 25 June 2009 Obama claimed that the proposed legislation "is a jobs bill" that has "been written carefully to address the concerns that may have been expressed in the past," that it is "balanced and sensible," and would cost Americans no more than a "postage stamp" a day. He concluded by stressing the importance of the vote in the House of Representatives and noted that "We have been talking about this issue for decades, and now is the time to finally act."

In the immediate aftermath of the House of Representative's vote in favour of the ACES Act President Obama made a number of speeches calling for the Senate to take action. In Remarks on Energy Legislation made following the House's vote on 26 June 2009 Obama expressed confidence that "… in the coming weeks and months the Senate will demonstrate the same commitment to addressing what is a tremendous challenge and an extraordinary opportunity." A day later he devoted his Weekly Address to the ACES Act and finished by "… urging the Senate to take this opportunity to come together and meet our obligations to our constituents, to our children, to God's creation, and to future generations." In further Remarks on Energy made on 29 June 2009 Obama noted that: "The House of Representatives came together to pass an extraordinary piece of legislation that will finally open the door to decreasing our dependence on foreign oil, preventing the worst consequences of climate change, and making clean energy the profitable kind of energy … In months to come, the Senate will take up its version of the energy bill, and I am confident that they too will choose to move this country forward." Obama recognised that passage of similar legislation in the Senate would be difficult and require months of negotiations and further compromises but urged the chamber's leadership to "seize the day" (Broder, 2009d).

A number of concerns and events conspired to delay the introduction of legislation similar to the ACES Act in the Senate. First, the narrow margin by which the ACES Act passed the House of Representatives, and the defection of a significant number of Democrats in the floor vote, provided a warning to the Senate's leadership that any bill would have to be carefully crafted to secure the support of all Democrats in the Senate. A warning of how difficult this task might prove came early in August 2009 when 10 Democratic Senators representing coal and heavy manufacturing states sent a letter to President Obama saying that they would not support any climate change legislation that failed to protect American industry from overseas competition (Broder, 2009e). They called for transition aid for energy-intensive industries, the negotiation of a strong international agreement on emissions, and increased funding for clean energy technology. Second, the two key Senators involved in the drafting of climate change legislation spent most of the August recess away from Washington, DC. Senator Barbara Boxer spent considerable time in California working on her 2010 re-election campaign while Senator John Kerry (D. MA) stayed in Massachusetts while he recovered from hip surgery. Although Senate staffers continued to work on draft legislation, the absence of the two Senators made it difficult to reconcile differences (Samuelsohn, 2009a). Third, the death of Senator Edward Kennedy (D. MA) on 25 August 2009 distracted the Senate and created political uncertainty. Kennedy's death raised questions about when a successor would take office and what that successor's position would be on climate change. Finally, the Senate became increasingly preoccupied with health care reform. The struggle to make progress on another of President Obama's key legislative programmes diverted time and attention away climate change.

Draft legislation based on the ACES Act was introduced in the Senate on 30 September 2009 by Senator Barbara Boxer and Senator John Kerry. The Clean Energy Jobs and American Power Act, also known as the Kerry-Boxer bill, contained a stronger cap on greenhouse gas emissions in the medium term than the ACES Act. Kerry-Boxer called for a 20 per cent reduction in emissions below 2005 levels by 2020 compared to the House measure's requirement of a 17 per cent reduction in the same time period. Both measures called for an 83 per cent reduction by 2050. In an effort to meet some of the concerns of Senators from coal and heavy industry states the legislation also proposed increased support for the nuclear energy industry, help for the clean energy economy, and a report on the efforts of countries such as China and India to reduce their greenhouse gas emissions. Details on how greenhouse gas emission allowances would be distributed under a cap-and-trade system were left out of the Kerry-Boxer bill in a deliberate attempt to provide the bill's authors with bargaining chips in negotiations with other Senators. Senators Boxer and Kerry intended to offer concessions on issues such as whether allowances would be auctioned or given away to secure support for the measure.

Senator Kerry stressed that the bill represented a "starting point" for discussions in remarks made just before its formal introduction (Samuelsohn, 2009b). The failure to provide details about how a cap-and-trade system would work, however, raised two obstacles to the successful conclusion of these discussions. First, the enormous impact that an emission allowance regime would have on economic interests meant that negotiations would be tough and time consuming as Senators sought to protect their constituents. Not only did the greater authority afforded to individual Senators pose challenges not present in the House of

Representatives, but doubts existed about the institutional capacity of the Senate to resolve differences on climate change at the same time as dealing with health care reform. Second, Republicans seized upon the lack of detail to criticise the bill. Senator James Inhofe (R. OK), for example, argued that without information on how emission allowances would be distributed "farmers, families and workers have no way of gauging how acutely they will be affected from job losses, higher electricity, food, and gasoline prices" (Black, 2009). No Republican Senator expressed support for the bill.

The prospect of Senate action on climate change sparked an intense lobbying effort by energy groups anxious to defend their interests (Krauss and Mouawad, 2009; Broder and Mouawad, 2009). These efforts encompassed a variety of tactics ranging from the production of studies that showed steep cost increases in petrol and electricity costs following the introduction of a cap-and-trade system to public demonstrations organised by industry sponsored front groups. Energy groups proved far from united in their approach to Kerry-Boxer, however, and important divisions were apparent between energy producers in particular. Conflict could be seen between the oil industry and the natural gas industry, the coal industry and the natural gas industry, within the electricity generating industry, and between the renewable energy industry and the other producers. Big Oil and parts of the coal industry preferred to defeat the legislation but other parts of the sector battled over the distribution of subsidies and benefits. How emission allowances would be allocated formed the primary focus of this lobbying effort. Energy groups sought to persuade senators to give away the allowances, or at least a high proportion of them, to help ease the cost of complying with the proposed law.

The result of this lobbying effort became apparent when Senator Boxer introduced a revised version (Chairman's Mark) of the Clean Energy Jobs and American Power Act on 23 October 2009. Included in the Bill were provisions to give away emission allowances to utilities and energy companies to cushion the impact on consumers. Free emission allowances were also given to energy-intensive manufacturing industries to ease the transition to a lower-carbon economy and to help them compete internationally with manufacturers based in countries without similar restrictions on greenhouse gas emissions. Primarily intended to secure the support of senators from heavy-manufacturing states, this provision also sought to address "carbon leakage" where industries re-located to countries without effective greenhouse gas emission limits. Benefits for the renewable and low-carbon energy industry included increased financing for low-carbon transportation programmes, the development of carbon capture and storage technology, measures to promote the use of natural gas, and the training of workers for the nuclear energy industry. In an effort to appease agricultural interests the new version of Kerry-Boxer provided additional support for farming and forestry, and greater help for rural communities.

In a press release accompanying the introduction of the Chairman's Mark of the Clean Energy Jobs and American Power Act Senator Boxer claimed that: "We've reached another milestone as we move to a clean energy future, creating millions of jobs and protecting our children from dangerous pollution. I look forward to the hearings and the markup as we move ahead to the next stage" (EPW, 2009a). To demonstrate her commitment to rapid progress Senator Boxer scheduled hearings before the Senate's Environment and Public Works Committee for the following week in a move that angered Republicans. Ranking Minority Member Senator Inhofe argued that: "It's not unreasonable to demand that a committee, prior to legislative hearings, would actually have the bill in question with adequate time for review and analysis" (Broder, 2009f). Concern about the proposed timetable for action also emerged at the hearings. Senator George V. Voinovich (R. OH), viewed as a possible supporter of climate change legislation, asked: "Why are we trying to jam this legislation now? Wouldn't it be smarter to take our time and do it right?" (Broder, 2009f). Republicans viewed the bill as too ambitious, too complex, and harmful to the economy (Eilperin, 2009). Even some Democrats expressed reservations about the bill and the rush to secure passage. Senator Max Baucus (D. MT), second-ranking member of the Committee, warned that "Montana, with our resource-based agriculture and tourism, cannot afford the unmitigated impacts of climate change. But we also cannot afford the unmitigated effects of climate change legislation," and counselled Senator Boxer to reach out to Republicans to build a "consensus here in this committee. If we don't, we risk wasting another month, another year, another Congress without taking a step forward into our future" (Broder, 2009f).

The problem facing Senator Boxer was that no Republicans wished to engage in a process to find a consensus that used Kerry-Boxer as a starting point. In an unusual parliamentary tactic the Committee's Republicans decided to boycott the mark-up of Kerry-Boxer forcing Senator Boxer to act in contravention of a rule that two members of the minority had to be present to report a bill. On 5 November 2009 the Democratic members of the Environment and Public Works Committee voted 11 to 1 to approve the bill without debate or amendment. Senator Baucus cast the only vote against. In a statement defending her action Senator Boxer claimed that: "A majority of the Committee believes that [Kerry-Boxer], and the efforts that will be built upon it, will move us away from foreign oil imports that cost Americans one billion dollars a day, it will protect our children from pollution, create millions of clean energy jobs, and stimulate billions of dollars of private investment" (EPW, 2009b). Senator Inhofe countered with a statement that expressed disappointment with "Chairman Boxer's decision to violate the rules and longstanding precedent of the committee" and warned that "her action signals the death knell for the Kerry Boxer bill" (EPW, 2009c). He argued that developing "all of America's vast energy resources" would be the best way to reduce dependence on foreign oil.

The conflict within the Environment and Public Works Committee over Kerry-Boxer revealed the obstacles that would have to be overcome to win Senate approval of the bill. First, the clear partisan division in the Committee suggested that President Obama and the Democratic leadership would have difficulty attracting Republican support for the measure. The Republican leadership viewed cap-and-trade as an issue that could harm the Democrats in the 2010 midterm elections and began to put pressure on potential supporters of Kerry-Boxer to vote against passage. Without the support of a few Republicans the prospects of success were negligible given the size of the Democratic majority and the likelihood of some Democrats voting against the measure. These concerns became even more pressing when the Republican Scott Brown won a special election in January 2010 to fill the vacancy caused by the death of Senator Kennedy. Brown's victory meant that Democrats no longer had a filibuster-proof majority in the Senate. Second, Senator Baucus's vote against Kerry-Boxer in the Committee raised the spectre of powerful opponents and confirmed that Democratic unity could not be taken for granted. Baucus served as chair of the Senate Finance Committee and was a senior member of the Senate Agriculture Committee, both of which would have a significant role in the shaping of any climate change legislation, and his concerns with Kerry-Boxer suggested that early floor action would be unlikely. Other moderate Democrats also began to express reservations about the bill. Senator Blanche Lincoln (D. AR) and Senator Byron Dorgan (D. ND) expressed concerns about the bill, while Senator Ben Nelson (D. NE) and Senator Mary Landrieu (D. LA) indicated that they would vote against passage.

Electoral concerns underpinned the unease that many Democrats felt towards Kerry-Boxer. Against a backdrop of a poorly performing economy and the struggle to pass health care reform Democrats began to fear that support for sweeping climate change legislation that included a cap-and-trade system would be an electoral liability. "We need to deal with the phenomena of global warming, but I think it's very difficult in the kind of economic circumstances we have right now," said Senator Evan Bayh (D. IN), who also viewed passage of any cap-and-trade measure as "unlikely" (Lerer, 2009). Scott Brown's victory in the Massachusetts special election in January 2010 served to reinforce Democratic fears about their electoral vulnerability. Many Democrat senators regarded Brown's victory as an indication of widespread voter discontent about the economy that posed a threat to the party's prospects in the midterm elections of November 2010, and began to call for climate change legislation to be dropped and replaced with a jobs bill that directly addressed the electorate's concerns. Persisting with the effort to enact climate change legislation, they feared, would provide Republican opponents with the opportunity to claim that they had spent time trying to pass a measure that would increase energy costs and lead to job losses.

The ebbing of support for climate change legislation in the Senate as the electoral season began in earnest indicated that the framing strategy employed by President Obama and congressional leaders to sell the measure had failed. Obama's efforts to portray the legislation as ushering in a "green economy" with millions of new "green" jobs, or as enhancing the country's "energy security" by reducing the dependence on foreign oil, proved too easy to counter with claims of jobs lost to international competitors and promises to exploit

America's oil, gas, and coal reserves. Republicans also proved adept at discrediting cap-and-trade as an effective policy tool. The weak economy, public mistrust of Wall Street, and a complex legislative proposal laden with favours to special interests allowed Republicans to label cap-and-trade as "cap and tax" or "tax-and-redistribution" scheme that would only lead to bigger government. "Economywide cap and trade died of what amounts to natural causes in Washington" declared Fred Krupp, president of the Environmental Defense Fund and a long-time advocate of the idea, "The term itself became too polarizing and too paralyzing in the effort to win over conservative Democrats and moderate Republicans to try to do something about climate change and our oil dependency" (Broder, 2010).

Negotiations to find a compromise bill acceptable to Democrats and moderate Republicans continued into the summer of 2010. In May 2010 Senator Kerry and Senator Lieberman (I. CN) released a discussion draft of legislation that sought to reduce greenhouse gas emissions, reduce oil imports, and create new jobs in the energy sector. Although the targets for greenhouse gas emission reductions were the same as in the ACES Act, the proposal contained plans to establish a cap-and-trade system for electric utility companies rather than an economy wide cap-and-trade system. Limited support for the proposal, however, meant that President Obama and Senate Majority Leader Harry Reid eventually accepted the inevitable. On 21 July 2010 Reid announced at a press conference that the Senate would not consider Kerry-Boxer, or any other comprehensive climate change bill, but would focus instead on responding to the Gulf of Mexico oil spill and tightening energy efficiency standards. "We know where we are," Reid stated, "We know that we don't have the votes" (Hulse and Herzenhorn, 2010). Carol Browner, the White House co-ordinator for Energy and Climate Policy, claimed at the same press conference that the Obama administration was disappointed that comprehensive climate change legislation would not be passed but accepted Senator Reid's decision. President Obama made no immediate comments on the failure to pass legislation.

The results of the midterm elections of November 2010 finally ended any prospects of climate change legislation during President Obama's first term in office. The Republican capture of the House of Representatives and the reduced majority in the Senate meant that discussions about how to reduce greenhouse gas emissions were replaced by attacks on the EPA's authority to regulate and reductions in funding. In April 2011 the House of Representatives passed the Energy Tax Prevention Act on a largely party-line vote (19 Democrats voted for the measure). This bill included provisions that would prevent the EPA from regulating greenhouse gases, and repeal all regulatory action already taken by the EPA related to climate change, including the endangerment finding, the "tailoring" rule, and the proposed New Source Performance Standards. Fred Upton (R. MI), the co-sponsor of the bill, argued that: "Our thoughtful, bipartisan solution reins in an EPA gone wild whose bureaucrats are oblivious to the nation's economic woes and soaring unemployment" (Broder, 2011). Senator Mitch McConnell (R. KY) offered similar language in a floor amendment to an unrelated bill but the Senate defeated the proposal on a 50–50 vote. Efforts to remove the funding for climate change constituted the second front in the attack on the EPA. In February 2011 the House passed a spending bill that would have removed the funding for the post of White House Co-ordinator for Energy and Climate Policy, prevented the EPA from regulating greenhouse gas emissions from stationary sources, and stopped US funding for the UN Intergovernmental Panel on Climate Change. Although this measure did not pass the Senate, the agreed spending bill for 2011–2012 did eliminate funding for the White House co-ordinator's post and a new centralised climate service within the National Oceanic and Atmospheric Administration (Vastag, 2011).

The removal of funding for the post of White House Co-ordinator for Energy and Climate Policy was symbolic as Carol Browner had resigned on 24 January 2011. Her departure resulted from a change in emphasis in the administration's policy on climate change that became evident in President Obama's State of the Union Address a day later. Obama made no mention of climate change in his Address but instead talked about initiatives to promote clean energy. This change of emphasis reflected both an acknowledgement of the fact that comprehensive climate change legislation would be impossible to push through Congress, and a recognition that the electorate did not view the issue as a priority. Obama began to promote smaller initiatives that promised investment opportunities and jobs rather than demanding passage of major new laws to re-shape the American economy, and relied upon executive action rather than law making to achieve as many of his objectives as possible. Even this more limited agenda failed to gain acceptance in Congress. President Obama let his exasperation show in his State of the Union Address on 24 January 2012 when he noted that: "The differences in this Chamber may be too deep right now to pass a comprehensive plan to fight climate change. But that's no reason why Congress shouldn't at least set a clean energy standard that creates a market for innovation." He continued to announce that in the absence of congressional action he would be setting such standards for the federal government.

Failure to pass comprehensive climate change legislation had two important consequences. First, laws establish a statement of purpose and policy framework that is more stable than presidential proclamations, statements, and rule-making. Laws are harder to change or repeal than executive actions. President Obama's efforts to combat climate change are vulnerable to reversal by a new president whereas a statute would be a more significant edifice. Second, the failure to pass legislation that set targets for greenhouse gas emission reductions undermined President Obama's ability to provide leadership in the international arena. Congressional opposition meant that Obama found it much more difficult to negotiate emission reduction targets with other countries and left him with little room for manoeuvre.

The International Arena

The election of Barack Obama promised a new era in American engagement with the international community on climate change. During the 2008 election campaign Obama had consistently criticised the Bush Administration for failing to provide leadership on the issue, and his early rhetoric promised a significant change in America's engagement with the international community. In his taped video-message to a bipartisan group of governors on 18 November 2008 he stressed that: "I look forward to working with all nations to meet this challenge [of climate change] in the coming years," and promised that "once I take office, you can be sure that the United States will once again engage vigorously in these [climate] negotiations, and help lead the world toward a new era of global cooperation on climate change." Further evidence of a willingness to play a leading role in international negotiations on climate change came when Obama announced his key energy and environment appointments on 15 December 2008. He noted that "the solution to global climate change must be global" and stated that "America will lead not just at the negotiating table—we will lead, as we always have, through innovation and discovery; through hard work and the pursuit of a common purpose." His Remarks on Energy made on 26 January 2009 reiterated this commitment to leadership on the issue. Obama promised that "we will make it clear to the world that America is ready to lead. To protect our climate and our collective security, we must call together a truly global coalition ... It is time for America to lead, because this moment of peril must be turned into one of progress."

President Obama made an effort to translate words into action when he convened a special meeting on climate change at the G8 Summit in L'Aquila, Italy, in July 2009. At this meeting the world's 17 leading industrialised countries (both developed and developing) agreed to take action to ensure that global temperatures would not rise by more than 2 degrees centigrade on pre-industrial levels. This agreement followed an earlier commitment by the G8 countries to reduce greenhouse gas emissions by 80 per cent by 2050. In remarks made at the summit on 9 July 2009 Obama hailed both decisions as "significant steps forward" towards agreeing a new framework on climate change at the forthcoming special UN session in Copenhagen in December 2009. He also promised that the United States would play a leading role in addressing climate change: "We ... have a historic responsibility to take the lead ... I know that in the past the United States has sometimes fallen short of meeting our responsibilities. So let me be clear: Those days are over." An analysis of the decisions made at the G8 Summit, however, reveals old problems lurking beneath the promises of change. First, the divide between developed and developing countries remained wide. Developing countries refused to agree to the targets wanted by developed countries to reduce emissions by 2050. Second, the divide between the United States and most European countries remained wide. President Obama felt unable to accept interim targets for emission reductions wanted by European countries because of a concern that such targets would undermine support for climate change legislation in the US Congress.

President Obama returned to the need for action to address climate change in a speech delivered to the UN Climate Change Summit in New York on 22 September 2009. He began by noting that "the threat from climate change is serious, it is urgent, and is growing" and warned that "our generation's response to this challenge will be judged by history." He acknowledged that the United States had been slow to act in the past but told delegates that "this is a new day. It is a new era." To provide evidence of this "new era" Obama proceeded to list the actions taken in the first eight months of his administration which he claimed had "done more to promote clean energy and reduce carbon pollution ... than at any other time in our history." He then turned to the need to agree a new climate treaty at Copenhagen. He recognised that the task would be difficult "in the midst of a global recession, where every nation's most immediate priority is reviving their economy and putting their people back to work" and stressed that "we also cannot allow the old divisions that have characterised the climate debate for so many years to block our progress." He stated that developed nations "still have a responsibility to lead" but that rapidly-developing nations ... must do their part as well," and both had a "responsibility to provide financial and technical assistance" to help the "poorest and most vulnerable" nations "adapt to the impacts of climate change and pursue low-carbon development."

Considerable hurdles stood in the way of the action called for in President Obama's speech. First, Congress's failure to enact legislation to cap greenhouse gas emissions in 2009 hamstrung American efforts to persuade other countries to commit to targets (Broder, 2009g). Obama eventually announced provisional targets consistent with those contained in the ACES Act passed by the House of Representatives in June 2009 in an effort to assert American leadership but doubts remained about the willingness of Congress to accept any limits on greenhouse gas emissions, particularly if not matched by other countries (Samuelsohn, 2009c; Broder, 2009h). Second, the gap between developed countries and developing countries remained vast and seemingly unbridgeable. Obama's efforts to engage with the problem of climate change and offer leadership had reduced the tensions between the European Union and the United States that had dominated international negotiations since the 1990s, but failed to persuade countries such as China and India to make significant changes in their positions (Bodansky, 2010). Both countries announced plans to reduce their carbon intensity over the next year but resolutely refused to accept targets for limiting actual emissions. Third, no agreement existed on who would pay for the financial and technical assistance promised to poor countries and how the money would be raised (Rosenthal, 2009). Poor countries indicated that they would not agree to any new treaty that failed to specify such financial details.

Agreement on a legally binding treaty to limit greenhouse gas emissions proved impossible to reach at Copenhagen in December 2009 (Christoff, 2010). In a speech at the plenary session of the conference on 18 December 2009 President Obama lectured delegates. He told them that "while the reality of climate change is not in doubt ... I think our ability to take collective action is in doubt right now, and it hangs in the balance." Continuing in this vein he stated: "we know the fault lines because we've been imprisoned by them for years. The time for talk is over ... Or we can choose delay, falling back into the same divisions that have stood in the way of action for years." Obama's speech, however, offered no way of breaking the deadlock at the conference (Goldenberg, 2009e). Rather than making concessions to break the log-jam he reiterated the American negotiating position. He insisted that all major economies must put forward national action plans that will reduce emissions, allow their emissions to be verified, and provide financial assistance to poor countries. The disappointment within the conference hall was palpable, but Obama had little room for manoeuvre as he lacked congressional support for action that committed the United States to action while exempting major competitors from similar obligations (Darwall, 2013). All that could be agreed at Copenhagen was a political accord that established a goal of keeping the increase in global temperatures below 2 °C, stated that developed countries would "commit to implement" emissions targets for 2020 while developing countries will "implement mitigation actions," and contained a number of other promises about verification and financing.

President Obama claimed that the accord was an "important breakthrough" at a press conference in Copenhagen on 18 December 2009. He argued that "for the first time in history, all of the world's major economies have come together to accept their responsibility to take action on the threat of climate change" but stressed that "going forward, we are going to have to build on momentum that we established in Copenhagen to ensure that international action to significantly reduce emissions is sustained and sufficient over time." Any momentum established at Copenhagen, however, quickly slowed. The failure of the Senate to enact climate change legislation in the summer of 2010 showed starkly the lack of support in the country for meaningful reductions in greenhouse gas emissions. Without American leadership even the modest promises made at Copenhagen began to unravel. At a meeting in Bonn, Germany, in August 2010 to prepare for COP-16 to be held in Cancun, Mexico, in November 2010, divisions between developed and developing countries, and the United States and China, became apparent once again. Developing countries claimed their emission reduction targets were voluntary while those of developed nations were compulsory, poor countries complained that the funds promised at Copenhagen were insufficient, and China objected to US proposals to monitor Chinese greenhouse gas emissions. A compromise agreement was eventually reached in Cancun that committed all countries to take action to control greenhouse gas emissions, but no legally binding targets were agreed. Divisions continued to plague negotiations at COP-17 in Durban, South Africa, in December 2011. All that could be agreed was a process to negotiate a successor to the Kyoto Protocol by 2015 to be implemented by 2020.

The difficulty experienced by President Obama in the international arena is testimony to the challenge of playing a "two-level game" involving consideration of both domestic politics and the demands of other countries (Putnam, 1988).

Although Obama's election in November 2008 appeared to promise a new era in international climate talks, old political realities soon surfaced to frustrate the potential for radical change. Continued domestic concerns about the economic cost of any treaty that imposed restrictions on greenhouse gas emissions, a loss of competitiveness if countries like China and India were exempted from similar requirements, and the possibility of ceding sovereignty to the United Nations, effectively left Obama with a "win set" of successful outcomes as limited as that available to President Clinton a decade earlier. Persuading other countries to accept an outcome that satisfied American domestic concerns proved impossible under such circumstances, particularly after the Senate's failure to act on the ACES Act. Obama had neither the leverage nor the room to manoeuvre necessary to engineer a new legally-binding climate treaty. He ensured that the United States took a seat at the table and engaged in negotiations but proved unable to move deliberations beyond promises of voluntary action and further talks.

The 2012 Elections

Debate about climate change figured more prominently in the Republican nomination campaign than the general election in 2012 with candidates using the issue both as a litmus test to attack each other's conservative credentials and to criticise President Obama's record in office (Bradley, 2011). While Republican candidates produced advertisements, made speeches, issued press releases, and mentioned climate change during the nomination process, the issue received far less attention once Mitt Romney had secured his party's blessing. The Republican Party Platform made only a fleeting reference to climate change, and the issue did not figure in any of the presidential debates. The Democratic Party Platform contained a small section on climate change, and President Obama referred to the issue in his stock campaign speeches, but proposals for action were left deliberately vague. The economy, health care, and foreign policy were the main battle grounds of the 2012 presidential elections and neither candidate wished to venture too far into other areas.

Familiar arguments about climate change appeared in the first Republican Candidate's Debate held in Simi Valley, California, on 7 September 2011. When asked if climate change was occurring Governor Rick Perry (R.TX) responded "Well, I do agree that there is—the science is—not settled on this. The idea that we would put America's economy at—at—at jeopardy based on a scientific theory that is not settled yet, to me, is just—is nonsense," and Rep. Michelle Bachmann (R. MN) took aim at President Obama's claims about "green jobs" by noting that: "The President told us he wanted to be like Spain when it came to green job creation, and yet Spain has one of the highest levels of unemployment." These remarks constituted the only significant effort to make substantive points about climate change during the nomination process. Far more common were efforts by the candidates to use climate change as a means of showing that their rivals lacked consistent positions on issues or were closet liberals. Rick Perry and Jon Huntsman attacked Romney for changing his position on climate change in late October and early November 2011, and Senator Rick Santorum (R. PA) claimed in February 2012 that Romney had supported cap-and-trade when Governor of Massachusetts. Romney and Bachman also attacked Newt Gingrich for appearing previously with Nancy Pelosi in a television advertisement to raise awareness of climate change. Romney stated at the Republican Candidates Debate held in Concord, New Hampshire, on 8 January 2012 that "You [Gingrich] had sat down with Nancy Pelosi and—and argued for—for a climate change bill." He followed this with a press release titled "Newt and Nancy's Loveseat." Revealed in these attacks is the extent to which climate change had become a key indicator of what it meant to be a conservative. Action to address climate change had become associated with liberal politicians such as Nancy Pelosi and Barack Obama and something for conservatives to oppose.

President Obama contrasted the denial and scepticism about climate change revealed in the Republican nomination process with the position of Senator John McCain in a number of campaign speeches. He observed that although he and McCain had some serious differences in 2008, they both agreed that climate change was a major problem and believed that cap-and-trade offered the best way to control greenhouse gas emissions. Unlike the 2008 campaign, however, Obama refused to specify what he proposed to do about climate change. In speeches in San Francisco on 16 February 2012 and Atherton, California, on 23 May 2012, he mentioned that re-election would mean "we can start tackling climate change in a serious way," but did not provide any details about what this meant. Mitt Romney sought to exploit Obama's reference to "serious" action with suggestions that he had a secret agenda that would harm the economy, but little came of such attacks. Climate change failed to ignite as a major issue in the 2012 campaign despite policy differences between the two candidates. While the Republican Party Platform published on 27 August 2012 included a statement calling "on Congress to take quick action to prohibit the EPA from moving forward with new greenhouse gas regulations that will harm the nation's economy and threaten millions of jobs over the next quarter century," for example, the Democratic Party Platform published on 3 September 2012 promised that we will "reduce emissions domestically—through regulation and market solutions." No discussion of these differences occurred in any of the presidential debates and no major references to climate change were made in speeches during the general election. Both candidates believed they could win the election by highlighting other issues.

A week after his election victory (14 November 2012), President Obama gave a news conference in which he was asked what action he would take to address climate change. His reply acknowledged that "for us to take on climate change in a serious way would involve making some tough political choices," and noted that the public remained concerned about economic growth and jobs, but expressed confidence that his new administration could "shape an agenda that says we can create jobs, advance growth, and make a serious dent in climate change and be an international leader, I think that's something that the American people would support." The problem confronting Obama, however, was that the 2012 election had not changed the "political stream" sufficiently to allow legislative action. The vague references to climate change in the campaign denied Obama the opportunity to claim a mandate for action, Republicans still controlled the House of Representatives, and the public showed only sporadic interest in the problem. These facts constrained the choices available to the administration. Following the pattern of the last two years of his first term in office Obama turned to executive power to advance his agenda and battled to defend his actions from congressional efforts to reverse them.

Unfinished Business

President Obama began his second term with a number of speeches that emphasised his determination to do something "serious" about climate change. In his Weekly Address on 5 January 2013 he stated that he looked forward to working with Congress to promote "our energy independence while protecting our planet from the harmful effects of climate change," and mentioned the issue again in his Weekly Address on 12 January 2013. His most important efforts to place climate change firmly on the political agenda, however, came in two major set-piece speeches. In his Inaugural Address on 21 January 2013 he declared: "We will respond to the threat of climate change, knowing that failure to do so would betray our children and future generations. Some may still deny the overwhelming judgement of science, but none can avoid the devastating impact of raging fires and crippling drought and more powerful storms." In his State of the Union Address on 12 February 2013 he warned again that: "... for the sake of our children and our future, we must do more to combat climate change." He continued to "urge this Congress to get together, pursue a bipartisan, market-based solution to climate change, like the one John McCain and Joe Lieberman worked on together a few years ago." He concluded with a shot across Congress's bows. "But if Congress won't act, I will," he warned, "I will direct my cabinet to come up with executive actions we can take, now and in the future, to reduce pollution, prepare our communities for the consequences of climate change, and speed the transition to more sustainable sources of energy."

Obama's exhortation for Congress to produce "a bipartisan, market-based solution to climate change" fell on deaf ears. No bills to establish a cap-and-trade system of regulating greenhouse gas emissions were introduced in the 113th Congress (2013–2014) let alone enacted. Obama had slightly more success in persuading Congress to consider legislation to adapt to the consequences of climate change. The Center for Climate and Energy Solutions identified 48 bills introduced in the 113th Congress intended to improve resilience to climate change compared to nine introduced in the previous Congress (C2ES, 2014). Only a measure to provide disaster relief in the aftermath of Hurricane Sandy was enacted of these 48 bills. The Disaster Relief Appropriation Act of 2013 (H.R. 152) provided additional funding for the Federal Emergency Management Agency's disaster relief fund, money to rebuild transportation systems, authorised the Army Corps of Engineers to conduct a study of flood risks of vulnerable coastal populations, and provided money to the Department of Housing and Urban Development for improving infrastructure resilience. The law made no specific mention of climate change.

With little sign of congressional action on climate change President Obama used a Proclamation designating the 22 April 2013 as Earth Day to give notice of his intent to take executive action. In Proclamation 8962 issued on 19 April 2013 he stated: "… I have called upon the Congress to pursue a bipartisan, market-based solution to climate change. In the meantime, I will direct my Cabinet to come up with executive action to reduce pollution, prepare our communities for the consequences of climate change, and speed our transition to sustainable energy." The result of the administration's search for executive tools to address climate change became public with the publication of "The President's Climate Action Plan" on 25 June 2013. The Plan outlined a number of initiatives that the administration would take over the remainder of Obama's term in office to reduce emissions of greenhouse gases, adapt to the consequences of climate change, and lead international efforts to address the problem. Some of these initiatives were relatively minor but others required major rule-making action. President Obama issued a "Memorandum on Power Sector Carbon Pollution Standards" on the same day that the Plan was announced, for example, that ordered the EPA to propose regulations for cutting greenhouse gas emissions for new power stations by September 2013 and existing power stations by June 2014. The "Memorandum" noted that power generation accounted for nearly 40 per cent of the country's carbon emissions.

President Obama defended his approach in a major speech at Georgetown University on 25 June 2013. He insisted that although he remained ready to work with legislators to produce a bipartisan, market-based law that addressed climate change, "this is a challenge that does not pause for partisan gridlock. It demands our attention now." Emphasising the urgency of the problem, Obama stated "… I don't have much patience for anyone who denies that this challenge is real. We don't have time for a meeting of the Flat Earth Society." He urged his audience to engage with the issue, seek to educate sceptics of the need for action, and let elected politicians know that attitudes towards climate change would be taken into account when deciding how to cast votes. Obama reiterated this message in a Weekly Address devoted exclusively to climate change on 29 June 2013. After outlining and defending his "Climate Action Plan," he stated "If you agree with me, I need you to act … Remind everyone who represents you, at every level of government, that there is no contradiction between a sound environment and a strong economy; and that sheltering future generations against the ravages of climate change is a prerequisite for your vote."

Republicans quickly recognised that President Obama's "Climate Action Plan" gave them a powerful issue to exploit in the midterm elections of 2014 (Gabriel, 2013). With Democrats vulnerable in energy-rich states such as Louisiana, Texas, West Virginia, and South Dakota, Republicans began airing advertisements warning of job losses and rising energy prices. In West Virginia the coal industry paid for bill boards that proclaimed the state as "Obama's No Jobs Zone." Rep. Shelley Moore Capito (R. WV) claimed that Obama's climate initiative "is a problem for every Democrat" in such areas (Gabriel, 2013). The prospect of electoral defeat led a number of Democrats representing coal mining states to distance themselves from the "Climate Action Plan." Rep. Nick Rahall (D. WV) defiantly declared that he was "profoundly disappointed" by President Obama's actions, and insisted that "I'm not ever, ever, ever going to back away from fighting for our coal miners" (Gabriel, 2013). Given the adverse electoral consequences likely to flow from his climate initiative, President Obama probably calculated that he would accept the loss of Democratic seats in Congress in return for action on one of his main priorities.

The EPA issued proposals to regulate greenhouse gas emissions from new power stations in September 2013 as ordered by President Obama. These would establish different New Source Performance Standards (NSPS) for gas-fuelled power stations and coal-fuelled power stations. New gas-fuelled power stations would be able to meet the standards relatively easily but those using coal would need to install carbon capture-and-storage technology to satisfy the proposed regulations. The EPA also met the June 2014 deadline set by President Obama for proposals to regulate greenhouse gas emissions from existing power stations. These would establish different target emission levels for each state, and give them considerable flexibility to determine how they would be met, with the overall goal of reducing emissions by 30 per cent from 2005 levels by 2030. Of the two sets of proposals those regulating existing power stations were the most significant. Virtually no new coal-fuelled power stations have been built or planned in the United States in recent years because of the low price of natural gas, whereas hundreds of existing plants still use coal (Gerrard, 2014).

President Obama used his "Weekly Address" on 31 May 2014 to prepare public opinion in advance of the EPA's announcement of the proposed rules for existing power stations. He warned his audience that "special interests and their allies in Congress will claim these guidelines will kill jobs and crush the economy. Let's face it, that's what they always say," but continued, "As President and as a parent, I refuse to condemn our children to a planet that's beyond fixing ... America will build the future: a future that's cleaner, more prosperous, and full of good jobs; a future where we can look our kids in the eye and tell them we did our part to leave them a safer, more stable world." The coal industry reacted to the proposed rules as Obama predicted. The American Coalition for Clean Coal Electricity, for example, argued that the regulations would hurt the economy and lead to power outages (Goldenberg, 2014). Environmentalist generally welcomed the proposed rules. Al Gore claimed that they were 'the most important step taken to combat the climate crisis in our country's history" (Goldenberg, 2014).

President Obama followed his "Memorandum on Power Sector Carbon Pollution Standards" with a number of other executive orders and memoranda designed to implement his Climate Action Plan. On 1 November 2013 he issued an executive order (EO 13653) that required the federal government to prepare new strategies for improving the country's "preparedness and resilience" to the consequences of climate change. He built upon this instruction in another executive order (EO 13677) issued on 23 September 2014 which required "climate-resilience" to be taken into account in all international development work involving the United States. Two other memoranda sought to give a boost to the production of clean energy. In a "Memorandum on Federal Leadership on Energy Management" issued on 5 December 2013, Obama set new goals for the use of renewable energy. The "Memorandum" required 20 per cent of all the energy consumed by the federal government to come from renewable sources by 2020. In a "Memorandum on Establishing a Quadrennial Energy Review" issued on 9 January 2014 Obama established a periodic review of the country's energy infrastructure.

Congressional Republicans, and a number of Democrats from coal-mining states, reacted strongly to President Obama's use of his executive power to address climate change. Senator Michael Enzi (R. WY) claimed that the EPA's intention to regulate greenhouse gas emissions from power stations would "… kill coal and its 800,000 jobs" and Rep. Nick Rahall (D. WV) declared: "I will oppose this rule as it will adversely affect coal miners and coal-mining communities" (Baker and Davenport, 2014). The Center for Climate and Energy Solutions identified 57 bills introduced in the 113th Congress (2013–2014) that proposed to block or limit the EPA's authority to regulate greenhouse gases under the Clean Air Act (C2ES, 2014). Five of these bills passed the House but none passed the Senate. Numerous other efforts were made to restrict or cut off funding for various other climate change activities undertaken by the US Government. The Consolidated Appropriations Act of 2014, which was signed into law on 17 January 2014, contained provisions to prohibit funding for the implementation of federal light bulb efficiency standards and to stop the administration from banning the funding of new overseas coal-fuelled power stations. Amendments to cut off funding for the National Climate Assessment and the IPCC report passed the House on several occasions but were not adopted by the Senate. The introduction of a number of bills to prohibit the regulation of greenhouse gases until the EPA Administrator certified that China, India and Russia had introduced similar restrictions also served notice of Congress's continued opposition to any action that appeared to place the American economy at a comparative disadvantage.

The "President's Climate Action Plan" announced on 25 June 2013 contained a commitment "to couple action at home with leadership internationally." President Obama's ability to offer leadership on the international stage, however, continued to be undermined by congressional opposition to any agreement that failed to treat other major emitters of greenhouse gases the same way. Knowing that President Obama would struggle to secure ratification of a new climate treaty left other countries sceptical of the administration's ability to deliver and created difficulties for American negotiators. These difficulties became evident at COP-19 held in Warsaw in November 2013 when developing countries sought to have the distinction between them and developed countries recognised again in discussions about mitigation obligations.

The conference ended with little agreement apart from a vague commitment to work towards a new draft of a climate treaty in time for discussion at COP-20 in December 2014 and a few measures to slow deforestation (Bernstein, 2013). In a speech at the United Nations Climate Summit in New York on 23 September 2014 President Obama sought to assert American leadership on the issue with a call to arms. "The alarm bells keep ringing. Our citizens keep marching. We have to answer the call," he told the Summit, and assured everyone that "there should be no question that the United States is stepping up to the plate. We recognise our role in creating the problem, and embrace our responsibility to combat it." Less than two months later Republican victories in the congressional midterm elections made the prospect of the Senate ratifying any climate change treaty as remote as ever.

Conclusion

Barack Obama entered office in January 2009 committed to addressing climate change. Over the course of the following six years he launched a number of policy initiatives to further this goal. Taking advantage of the opportunity provided by the need to respond to the country's economic woes he used the ARRA to boost government support for a range of clean and alternative energy initiatives. His proposals for a law establishing a comprehensive cap-and-trade system to regulate greenhouse gas emissions passed the House of Representatives but eventually floundered in the Senate. Subsequent calls for climate change legislation fell on deaf ears as electoral results gave Republicans a majority in the House in 2010, the Senate in 2014, and made a significant number of Democrats wary of supporting measures that could be characterised by opponents as harming the economy and costing jobs. Lacking congressional support for his policy preferences Obama made increasing use of executive orders, presidential memoranda, proclamations, and his rule-making powers to further his agenda. These culminated in June 2014 with the EPA's publication of rules to regulate greenhouse gas emissions from existing power stations. On the international stage Obama brought the United States back to the negotiating table, but found securing agreement on a successor to the Kyoto Protocol difficult given long-standing congressional opposition to legally binding emission targets and the need to accommodate the demands of other countries. The announcement of an agreement between the United States and China in November 2014 setting goals for future greenhouse gas emissions indicates that the Obama Administration has accepted that the Senate will not ratify a climate change treaty in the near future, and is pursuing a strategy of bilateral "politically-binding" agreements that do not require ratification to pursue its objectives (Landler, 2014; Davenport, 2014).

President Obama's use of his executive powers to achieve his policy objectives points to familiar problems in addressing climate change. Despite overwhelming scientific evidence that anthropogenic climate change is occurring and will have profound consequences for the United States, the American public remains unengaged with the issue, powerful economic interests opposed to regulation have the resources to influence government decisions, and most Republicans continue to dismiss the need for action. Faced with such unfavourable circumstances, Obama used his presidential power to bring about a radical change in policy that contrasts with the incrementalism of previous climate change policy. The question is whether this radical change in policy will stick. Executive actions can be reversed by future presidents with different priorities, hamstrung by congressional funding decisions, and bogged down by litigation.

Conclusions

Over the last forty years or so the US government has produced a range of policies to address climate change. Laws, regulatory action, and court rulings have led to advances in climate science, action to reduce levels of greenhouse gas emissions, and efforts to prepare for the potential consequences of climate change. This record gives the lie to claims that inaction has characterised the US government's response to climate change. The US government has not always placed a high priority on dealing with climate change; it has failed to take a consistent approach to the problem, it has pursued different approaches at different speeds, and, undoubtedly, it has disappointed environmentalists and scientists who believe more concerted action has been needed. But a careful examination of the evidence reveals a number of policy actions taken since the 1970s that have been designed to investigate, mitigate, and adapt to climate change.

The broad development of the US government's climate change policies has been driven by the nature and interplay of the "problem," "policies," and "politics" streams (Kingdon, 2011). Policy outputs have reflected changing levels and acceptance of scientific knowledge, the acceptability of different solutions at particular times, and a political environment that has rarely been amenable to major shifts in policy. Open "policy windows" have been few and far between giving rise to long periods of policy equilibrium, during which policy has developed incrementally. Efforts to bring about radical changes in policy have proved politically divisive and have usually been short-lived. Conflict has been a defining feature of climate change policymaking, with no political consensus about the problem or what to do about it.

Disputes over the evidentiary basis of climate change, and uncertainty about the consequences of a warming planet, have meant that the problem has struggled to gain attention compared to other apparently more urgent concerns such as the economy, health care, and war. Although opinion polls suggest that a majority of Americans believe that climate change is occurring, the percentage which believes that human activity is the cause of any warming is less convincing, and most regard other environmental problems as more pressing. Climate change has lacked the "alarmed discovery" moment that has produced a number of major environmental statutes (Downs, 1972). There has been no equivalent of *Silent Spring*, tankers spilling oil on Californian beaches, deadly smogs in American cities, the Cuyahoga River catching fire, or hazardous waste seeping into the basements of houses in Love Canal to prompt dramatic policy action (see Smith, 1992, 15–16). Scientists and others may point to devastating hurricanes, lengthy periods of drought, raging forest fires and increased flooding as harbingers of climatic change, but the causal connection is not obvious enough for these to act as catalysts or triggers for radical action. The gradual accumulation of evidence about climate change, rather, has lent itself to an incremental approach to dealing with the problem.

Two important features of the debate about climate change have shaped the policy responses of the US government. First, climate change has been viewed primarily as an energy, rather than an environmental, issue virtually from the outset. This has meant that concerns about energy costs, supply, and security have usually dominated debates about policy, while fears about environmental damage have taken second place. The consequences of this particular farming are significant as the supply of cheap energy is seen by many as a cornerstone of the American way of life. "Don't walk, drive" could almost be a *leitmotif* for large parts of the population. To garner public support, proponents of action to address climate change have had to counter fears that any such action would lead to increases in energy prices. This has not proved an easy task. Second, an overt concern about the economic costs of dealing with climate change has been evident in debates in the United States since the early 1990s, whereas many European countries have glossed over or ignored such concerns in their policy deliberations. The consequences of this particular framing are similar to that flowing from viewing climate change as an energy issue. Proponents of government action have had to show that dealing with climate change will not cause economic hardship or give the country's competitors a comparative advantage. This has often proved difficult.

Advocates of action to address climate change have made a number of efforts over the last four decades to counter the arguments of opponents and even frame the issue in alternative ways. Presidents Clinton and Obama made a number of speeches, for example, in which they challenged the idea that the regulation of greenhouse gas emissions would lead to substantially higher energy prices and undermine the competitiveness of American industries. They argued instead that such action would boost jobs in the alternative energy industry, enhance energy security, and mean that the country would be well-positioned in a post-fossil fuel world. The difficulty with these arguments is that job creation and energy security could also be promoted by exploiting America's substantial supplies of fossil fuels. In the phrase popularised by Sarah Palin in the vice presidential debate held on 2 October 2008, the answer to such concerns might simply be "Drill, baby, drill." Both presidents also tried to move debate way from energy and the economy by stressing moral and health considerations in their speeches. President Obama's later speeches on the issue increasingly mentioned obligations to future generations and the health of children.

An emphasis on research, voluntary action, and technological fixes has dominated the US government's approach to dealing with climate change. These policy options not only fit with the country's dominant political and cultural values, but also often have a strong distributive element. Government spending, subsidies, or tax breaks to foster research or promote the development of new technologies provide short-term benefits to particular economic interests and constituencies that can boost the re-election prospects of members of Congress. "Pork-barrelling" of this sort is a long-established tradition in policy areas ranging from public works to defence spending to education, and can also be found in approaches to dealing with environmental problems such as hazardous waste sites (Evans, 2004; Lyons, 1999). In contrast, little enthusiasm for regulatory action to stabilise or reduce greenhouse gas emissions has been evident. This lack of enthusiasm is a product of both the strong anti-regulatory mood evident in the in country over the last 40 years and the political power of those economic interests that would be harmed by such regulations. The fate of the American Clean Energy and Security (ACES) Act of 2009 illustrates the difficulties of securing legislative support for regulatory action. First, the bill's sponsors had to transform the proposal into a massive distributive package in order to "buy" the votes of members of the House of Representatives. The bill passed by the House ended up being a prime example of environmental "pork-barreling" dripping with concessions to vested and constituency interests. Second, the bill failed to pass the Senate. The support of senators evaporated in the face of lobbying by the fossil fuel industry and concern about electoral prospects in the midterm elections of 2010.

The fate of the ACES Act illustrates Kingdon's observation that "Ideas, proposals, or issues may rise into or fall from favor from time to time" (2011, 141). "Cap-and-trade" as a means of dealing with air pollution gained favour among policymakers in the 1990s, figured largely in debates about climate change policy during the 2000s, but had passed its sell-by date at the federal level by 2010. Enthusiasm for other solutions such as carbon taxes followed a similar pattern. Carbon taxes fell out of favour following Congress's rejection of the Clinton Administration's proposals in 1993, remained moribund for nearly two decades, but have recently shown signs of life as "cap-and-trade" has fallen out of favour. As Kingdon notes, ideas rarely become extinct, but often lie dormant for periods of time waiting to be re-discovered. Interest in "cap-and-trade" at the federal level, for example, might grow again if California's use of the approach which started in 2013 works well and has few adverse consequences. States have often served as "policy laboratories" where different approaches to tackling problems can be tested before adoption at the federal level, and as the most populous and prosperous state in the country, California has often driven the development of national policy, particularly environmental policy (Selin and VanDeveer, 2007; Lutsey and Sperling, 2007; Rabe, 2004).

Legislative stalemate at the federal level has meant that old laws have sometimes been pressed into new uses to address climate change. President Obama's use of the Clean Air Act of 1990 to regulate greenhouse gas emissions is an obvious example of this. The Clean Air Act was not intended to address climate change but key provisions have proved adaptable enough for that purpose. Suggestions have also been made that the Clean Water Act of 1972 and the Endangered Species Act of 1973 could be used to tackle climate change (Adler, 2011). Evidence that increased levels of atmospheric carbon dioxide has contributed to ocean acidification raises the possibility of using the Clean Water

Act to regulate greenhouse gas emissions while the threat to wildlife habitat caused by climate change offers an opportunity to regulate using the Endangered Species Act. The problem with such approaches to policymaking, however, is the tenuous nature of rule-making (Klyza and Sousa, 2008, 106–8). Administrative action can be reversed by a new president, challenged in the courts, and blocked by Congress using its power of the purse. Clear statements of purpose in a new law offer a more stable way of addressing climate change.

Broad changes in the "problem," "policy," and "politics" streams have clearly shaped the US government's general approach to dealing with climate change, but do not explain the precise timing and content of policy outputs. Why specific legislative or regulatory actions were taken at particular times depends not only on the content and interplay of the various "streams," but also the actions of individuals and the vagaries of chance. A key element of the story of the US government's climate change policy is the important role played by policy entrepreneurs. Various presidents and legislators have played a central part in raising awareness of the problem, advancing solutions, and negotiating agreements. This has never been an easy task. The information and transaction costs of fashioning a policy response to climate change are extremely high. Meeting these costs takes resources and skills that are not equally shared. Presidents possess personal and executive authority that helps facilitate action, but potential policy entrepreneurs in Congress may lack the authority and resources needed to produce a policy output. Party leaders and committee chairs possess greater authority and resources than rank-and-file members but even they may struggle to overcome the numerous obstacles that characterise the legislative process. Overcoming such obstacles requires both entrepreneurial skill and often an element of luck. A policy entrepreneur needs to recognise opportunities for action and be able to broker deals. The "horse-trading" often involved in such negotiations help explain the preference for distributive policy found in Congress. "Pork-barrelling" provides a means to overcome the high transaction costs involved in climate change policymaking.

The high costs of legislative production mean that incrementalism has characterised the development of the US government's climate change policy for most of the last 40 years. Radical departures in approach have been rare and have been difficult to make stick. The Clinton Administration's decision to accept legally binding greenhouse gas emissions targets in the Kyoto Protocol, for example, was reversed by the Bush Administration within weeks of taking office. The Obama Administration's decision to regulate greenhouse gas emissions from power stations faces similar challenges in the future given the configuration of the contemporary "politics" stream. Not only have congressional Republicans promised to curb the regulatory authority of the Environmental Protection Agency (EPA) and reduce funding for a number of existing climate change programmes, but the legal basis of the Administration's rule-making has been challenged in the federal courts. The election of a Republican president in 2016 would likely end any prospect of further radical action and signal a return to the policy equilibrium of the past decades.

Kingdon's conceptual framework suggests that the prognosis for a radical change in the US government's climate change policy is not good. Although improvements in scientific knowledge will no doubt continue to provide greater understanding and evidence of climate change and its consequences, and policy experts will carry on refining the solutions available to address the issue and learn from the "experiments" conducted by states like California, these changes in the "problem" and "policies" streams will probably not be sufficient to overcome many of the stumbling blocks to action that have existed for the last 40 years. Climate change is likely to continue to be regarded as a distant problem with uncertain consequences which will require very expensive action to stop, let alone reverse. Changes in the "politics" stream conducive to prompting radical action on climate change are even less likely to occur. The public shows little sign of embracing the large governmental programmes associated with regulating greenhouse gases, the fossil fuel industry will continue to campaign against the need for action, and members of Congress have provided no indication that they are likely to support major legislation on the issue. The deep partisan divisions that have come to characterise congressional climate change politics show little sign of diminishing. Federal courts may also frustrate presidential action if cases challenging rule-making appear before conservative judges appointed by previous Republican presidents. In the policy environment likely to dominate the immediate future, policy change is likely to be at the margins, with perhaps an increased emphasis on adaptation, rather than radical.

Bibliography

All presidential speeches, statements, press releases, candidate campaign speeches, and party platforms referred to in this book can be accessed online at The American Presidency Project (www.presidency.ucsb.edu). I have provided the date and title of these speeches and statements in the text to enable readers to find them using this database.

Abramowitz, Michael and Steven Mufson (2007) "Papers Detail Industry's Role in Cheney's Energy Report," *The New York Times*, 18 July.

Adler, Jonathan H. (2011) "Heat Expands All Things: The Proliferation of Greenhouse Gas Regulation Under the Obama Administration," *Harvard Journal of Law and Public Policy*, 34: 421–52.

Agrawala, Shardul (1998) "Context and Early Origins of the Intergovernmental Panel on Climate Change," *Climatic Change*, 39: 605–20.

Alexander, Tom (1974) "Ominous Changes in the World's Weather," *Fortune*, February.

Andrews, Edmund L. (2001) "Bush Angers Europe by Eroding Pact on Warming," *The New York Times*, 1 April.

Antilla, Lisa (2005) "Climate of Scepticism: US Newspaper Coverage of the Science of Climate Change," *Global Environmental Change*, 5: 338–52.

Baer, Hans A. (2012) *Global Capitalism and Climate Change* (Lanham, MD: Rowman and Littlefield).

Baker, Peter and Coral Davenport (2014) "Using Executive Power, Obama Begins His Last Big Push on Climate Policy," *The New York Times*, 31 May.

Ball, Tim (2014) *The Deliberate Corruption of Climate Science* (Seattle: Stairway Press).

Bang, Guri (2010) "Energy Security and Climate Change Concerns: Triggers for Energy Policy Change in the United States," *Energy Policy*, 38: 1645–53.

Bang, Guri, Andreas Tjernshaugen, and Steinar Andressen (2005) "Future US Climate Policy: International Re-Engagement? *International Studies Perspectives*, 6: 285–303.

Barnum, Alex (1997) "US Works to Bridge Gap with the Third World," *The San Francisco Chronicle*, 3 December.

Barringer, Felicity (2012) "For New Generation of Power Plants, a New Emission Rule from the EPA," *The New York Times*, 27 March.

Beder, Sharon (2002) "Casting Doubt and Undermining Action," *Pacific Ecologist*, 1: 42–9.

Begley, Sharen (2007) "The Truth About Denial," *Newsweek*, 13 August.

Bennet, James (1997) "Warm Globe, Hot Politics: For Clinton and Gore, Fight Looms in the Senate," *The New York Times*, 11 December.

Bernstein, Lenny (2013) "Warsaw Climate Conference Produces Little Agreement," *The Washington Post*, 22 November.

Besel, Richard D. (2007) "Communicating Climate Change: Climate Rhetorics and Discursive Tipping Points in the United States Global Warming Science and Public Policy," Unpublished PhD thesis, University of Illinois at Champaign.

Black, Richard (2009) "US Bill Crucial For Climate Talks," BBC News, 30 September. www.news.bbc.co.uk/1/hi/sci/tech/8283655.stm. Accessed 5 March 2012.

——— (2011) "Polar ice loss quickens, raising seas," BBC News, 9 March. www.bbc.co.uk/news/science-environment-12687272. Accessed 27 October 2011.

Bodansky, Daniel M. (1993) "The United Nations Framework Convention on Climate Change: A Commentary," *Yale Journal of International Law*, 18: 451–558.

——— (2010) "The Copenhagen Climate Change Conference: A Postmortem," *The American Journal of International Law*, 104: 230–40.

Boehmer-Christiansen, Sonja (1994a) "Global Climate Protection Policy: the Limits of Scientific Advice, Part 1." *Global Environmental Change*, 4: 140–59.

——— (1994b) "Global Climate Protection Policy: the Limits of Scientific Advice, Part 2." *Global Environmental Change*, 4: 185–200.

Bolin, Bert (2007) *A History of the Science and Politics of Climate Change* (Cambridge: Cambridge University Press).

Bomberg, Elizabeth and Betsy Super (2009) "The 2008 US Presidential Election: Obama and the Environment," *Environmental Politics*, 18(3): 424–30.

Booker, Christopher (2010) *The Real Global Warming Disaster* (London: Continuum).

Borick, Christopher P., Erick Lachapelle, and Barry G. Rabe (2011) "Climate Compared: Public Opinion on Climate Change in the United States and Canada," *Issues in Governance Studies*, Brookings Institution, April.

Boykoff, Maxwell (2010) "U.S. Climate Coverage in the '00s," *Extra!*, Fairness & Accuracy In Reporting (FAIR), February. www.fair.org. Accessed 17 November 2011.

——— (2011) *Who Speaks for the Climate?* (Cambridge: Cambridge University Press).

Bowen, Alex and Samuel Frankhauser (2011) "The Green Growth Narrative: Paradigm Shift Or Just Spin?" *Global Environmental Change*, 21: 1157–9.

Bowen, Mark (2009) *Censoring Science* (New York: Dutton).

Boykoff, Maxwell and Jules M. Boykoff (2007) "Climate Change and Journalistic Norms: a Case Study of US Mass-Media Coverage," *Geoforum*, 38(6): 1190–204.

Bradley, Raymond S. (2011a) "Global Warming Is a Litmus Test for US Republicans," *The Guardian*, 3 August.

——— (2011b) *Global Warming and Political Intimidation* (Amherst, MA: University of Massachusetts Press).

Brewer, Paul R. (2012) "Polarisation in the USA: Climate Change, Party Politics, and Public Opinion in the Obama Era," *European Political Science*, 11: 7–17.

Brewer, Thomas L. (2014) *The United States in a Warming World: The Political Economy of Government, Business and Political Responses to Climate Change* (Cambridge: Cambridge University Press).

Brinkman, Robert and Sandra Jo Garren (2011) "Synthesis of Climate Change Policy in Judicial, Executive, and Legislative Branches of Government," *PORTAL*, 8(3): 1–26.

Broder, John M. (2007) "Governors Join In Creating Regional Pacts on Climate Change," *The New York Times*, 1 July.

——— (2008) "Democrats Oust Longtime Leader of House Panel," *The New York Times*, 21 November.

——— (2009a) "From a Theory to a Consensus on Emissions," *The New York Times*, 17 May.

——— (2009b) "US to Issue Tougher Fuel Standards for Automobiles," *The New York Times*, 19 May.

——— (2009c) "Obama Opposes Trade Sanctions in Climate Bill," *The New York Times*, 29 June.

——— (2009d) "With Something for Everyone, Climate Bill Passed," *The New York Times*, 1 July.

——— (2009e) "Climate Bill is Threatened by Senators," *The New York Times*, 7 August.

——— (2009f) "Senate Global Warming Bill Is Seeking to Cushion the Impact on Industry," *The New York Times*, 25 October.

——— (2009g) "Obama Hobbled in Fight Against Global Warming," *The New York Times*, 16 November.

——— (2009h) "Obama to Go to Copenhagen With Emissions Target," *The New York Times*, 26 November.

——— (2010) "'Cap and Trade' Loses Its Standing as Energy Policy of Choice," *The New York Times*, 25 March.

——— (2011) "House Votes to Bar EPA From Regulating Indistrial Emissions," *The New York Times*, 7 April.

Broder, John M. and Jad Mouawad (2009) "Energy Firms Deeply Split on Bill to Battle Climate Change," *The New York Times*, 19 October.

Brune, Michael (2011) "How President Obama Can Reclaim His Green Cred" *The Los Angeles Times*, 11 July.

Bruni, Frank (2000) "Bush, in Energy Plan, Endorses New U.S. Drilling to Curb Prices," *The New York Times*, 30 September.

Bryner, Gary (2008) "Failure and Opportunity: Environmental Groups in US Climate Change Policy," *Environmental Politics*, 17: 319–36.

Burkeman, Oliver (2003) "Memo exposes Bush's New Green Strategy," *The Guardian*, 4 March.

Burns, Wil C. and Andrew L. Strauss (eds) (2013) *Climate Change Geoengineering* (Cambridge: Cambridge University Press).

Byrne, John, Kristen Hughes, Wilson Rickerson and Lado Kurdgelashvili (2007) "American Policy Conflict in the Greenhouse: Divergent Trends in Federal, Regional, State, and Local Green Energy and Climate Change Policy," *Energy Policy*, 35: 4555–73.

C2ES (2014) "Climate Debate in Congress," Center for Climate and Energy Solutions. www.c2es.org/print/federal/congress. Accessed 13 October 2014.

Cairney, Paul (2012) *Understanding Public Policy* (Basingstoke: Palgrave Macmillan).

Cannon, Jonathan and Jonathan Riehl (2004) "Presidential Greenspeak: How Presidents Talk about the Environment and What It Means," *Stanford Environmental Law Journal*, 23: 195.

Carlarne, Cinnamin Pinon (2010) *Climate Change Law and Policy* (Oxford: Oxford University Press).

Carter, Robert M. (2010) *Climate: the counter-consensus* (London: Stacey International).

Cass, Loren C. (2006) *The Failures of American and European Climate Policy* (Albany, NY: SUNY Press).

Chait, Jonathan (2013) "Obama Might Actually Be the Environment President," *New York Magazine*, 5 May.

Christiansen, Atle C. (2003) "Convergence or Divergence? Status and Prospects for US Climate Strategy," *Climate Policy*, 3: 343–58.

Christiansen, Gale E. (1999) *Greenhouse: The 200-Year Story of Global Warming* (New York: Penguin Books).

Christoff, Peter (2010) "Cold Climate in Copenhagen: China and the United States at COP15," *Environmental Politics*, 19: 637–56.

Christoff, Peter and Robyn Eckersley (2011) "Comparing State Responses" in John S. Dryzek, Richard B. Norgaard, and David Schlosberg (eds) *The Oxford Handbook of Climate Change and Society* (Oxford: Oxford University Press).

Clarke, Chris (2002) "Bush Bashes Climate Report," *Earth Island Journal*, Fall. www.earthisland.org. Accessed 9 August 2013.

Clinton, Hillary (2009) "Appointment of Special Envoy for Climate Change Todd Stern," US Department of State, 26 January. www.state.gov/secretary/rm/2009a/01/115409.htm. Accessed 23 March 2012.

CNN (2001) "Groups Blast Bush for Reversing Position on Emissions Reductions," 15 March. www.edition.cnn.com. Accessed 15 May 2013.

Cohen, Michael, James March, and Johan Olsen (1972) "A Garbage Can Model of Organizational Choice," *Administrative Science Quarterly*, 17: 1–25.

Condon, Bradley J. and Tapen Sinha (2013) *The Role of Climate Change in Global Economic Governance* (Oxford: Oxford University Press).

Cushman, John H. (1996) "U.S. Will Seek Pact on Global Warming," *The New York Times*, 17 July.

——— (1998a) "Washington Skirmishes over Treaty on Warming," *The New York Times*, 11 November.

———— (1998b) "US signs a Pact to Reduce Gases Tied to Warming," *The New York Times*, 13 November.

Daniels, Brigham (2011) "Addressing Global Climate Change in an Age of Political Climate Change," *Brigham Young University Law Review*, 6: 1899–935.

Darwall, Rupert (2013) *The Age of Global Warming: A History* (London: Quartet).

Davenport, Coral (2014) "Obama Pursuing Climate Accord in Lieu of Treaty," *The New York Times*, 26 August.

Deland, Michael R. (1991) *America's Climate Change Strategy: An Agenda for Action* (Washington, DC: The White House).

Demeritt, David (2001) "The Construction of Global Warming and the Politics of Science," *Annals of the Association of American Geographers*, 91: 307–37.

Dewar, Helen (1997) "Senate Advises Against Emissions Treaty That Lets Developing Countries Pollute," *The Washington Post*, 26 July.

Dobriansky, Paula J. (2001a) "Statement to the Resumed Session of the Sixth Conference of the Parties (COP-6) to the UN Framework Convention on Climate Change," Bonn, Germany, 19 July. www.2001-2009.state.gov/g/rls/rm/2001/4152.html. Accessed 30 March 2015.

———— (2001b) "Closing Statement to the Seventh Session of the Conference of the Parties (COP-7) to the UN Framework Convention on Climate Change," Marrakesh, Morocco, 9 November. www.2002-2009.state.gov/g/oes/rls/rm/6050.html. Accessed 30 March 2015.

Dowie, Mark (1993) *Losing Ground* (Cambridge, MA: MIT Press).

Downie, Christian (2014) *The Politics of Climate Change Negotiations* (Cheltenham: Edward Elgar).

Downs, Anthony (1972) "Up and Down With Ecology: The 'Issue-Attention Cycle,'" *The Public Interest*, 28: 38–50.

Drew, Elizabeth (1997) *Showdown* (New York: Touchstone).

Driesen, David M. (2010a) "Introduction" in David M. Driesen (ed) *Economic Thought and US Climate Change Policy* (Cambridge, MA: The MIT Press).

———— (2010b) "Neoliberal Instrument Choice" in David M. Driesen (ed) *Economic Thought and US Climate Change Policy* (Cambridge, MA: MIT Press).

Dryzek, John S. (1990) *Discursive Democracy* (Cambridge: Cambridge University Press).

Duggan, Andrew (2014) "Americans Most Likely To Say Global Warming is Exaggerated," Gallup Organization, 17 March.

Dunlap, Riley E. (2008) "Climate-Change Views: Republican-Democratic Gaps Expand," Gallup Poll, 29 May.

Dunlap, Riley E. and Aaron M. McCright (2011) "Organized Climate Change Denial" in John S. Dryzek, Richard B. Norgaard, and David Schlosberg (eds) *The Oxford Handbook of Climate Change and Society* (Oxford: Oxford University Press).

EIA (US Energy Information Administration) (1998) "Impacts of the Kyoto Protocol on US Energy Markets and Economic Activity," Energy Information Administration, October.

Eilperin, Juliet (2009) "Economics of Climate Change in Forefront," *The Washington Post*, 28 October.

Engel, Kirsten H. (2010) "Courts and Climate Policy: Now and in the Future" in Barry G. Rabe (ed.) *Greenhouse Governance* (Washington, DC: Brookings).

Erikson, Robert S. and Kent L. Tedin (2014) *American Public Opinion*, 9th edition (New York: Pearson).

EU (2001) "EU Reaction to the Speech by US President Bush On Climate Change," IP/01/821, Brussels, 12 June.

Evans, Diane (2004) *Greasing the Wheels* (Cambridge: Cambridge University Press).

Evans, Rowland and Robert Novak (1981) *The Reagan Revolution* (Okemos, MI: Dalton).

Farber, Daniel A. (2011) "Issues of Scale in Climate Governance" in John S. Dryzek, Richard B. Norgaard, and David Schlosberg (eds) *The Oxford Handbook of Climate Change and Society* (Oxford: Oxford University Press).

Fialka, John (1997) "Global-Warming Treaty is Approved," *The Wall Street Journal*, 11 December.

Fisher, Dana R.(2004) *National Governance and the Global Climate Change Regime* (Lanham: Rowan and Littlefield).

Fisher, Dana R., Philip Leifeld, and Yoko Iwaki (2013) "Mapping the Ideological Networks of American Climate Politics," *Climate Change*, 116: 523–45.

Fleming, James Rodger (1998) *Historical Perspectives on Climate Change* (Oxford, Oxford University Press).

Fletcher, Amy L. (2009) "Clearing the Air: the Contribution of Frame Analysis to Understanding Climate Policy in the United States," *Environmental Politics*, 18: 800–816.

Fone, Joe (2013) *Climate Change: Natural or Manmade* (London: Stacey International).

Fuller, Thomas and Andrew C. Revkin (2007) "Climate Plan Looks Beyond Bush's Tenure," *The New York Times*, 16 December.

Gabriel, Trip (2013) "GOP Sees Opportunity for Election Gains in Obama's Climate Change Policy," *The New York Times*, 1 July.

Gallup (1999) "Environmental Concern Wanes in 1999 Earth Day Poll," Gallup Organization, 22 April 1999, www.gallup.com. Accessed 5 April 2013.

———— (2000a) "Americans Are Environmentally Friendly, But Issue Not Seen as an Urgent Problem," Gallup Organization, 17 April. www.gallup.com. Accessed 5 April 2013.

———— (2000b) "Environment Not Highest-Priority Issue This Year," Gallup Organization, 25 September. www.gallup.com. Accessed 1 May 2013.

———— (2011a) "Gallup Poll Social Series: Environment," Gallup New Service, 3–6 March.

———— (2011b) "Worldwide, Blame for Climate Change Falls on Humans," Gallup Poll, 22 April.

Gelb, Leslie (1991) "Sununu v. Scientists," *The New York Times*, 10 February.

Gelbspan, Ross (1997) *The Heat is On* (Reading, MA: Addison-Wesley).

———— (2004) *Boiling Point* (New York: Basic Books).

———— (2005) "Hurricane Katrina's Real Name," *The Boston Globe*, 30 August.

Gerrard, Michael B. (ed.) (2008) *Global Climate Change and US Law* (Chicago: American Bar Association).

———— (2014) "President Obama Tackles Climate Change Without Congress," *Trends*, 45: 2–5.

Gerson, Michael (2012) "Climate and the Culture War," *The Washington Post*, 17 January.

Gerstenzang, James (1995) "GOP Clouds the Future of Environmental Protection," *Los Angeles Times*, 24 December.

Gillis, Justin and Leslie Kaufman (2012) "Leak Offers Glimpse of Campaign Against Climate Science," *The New York Times*, 15 February.

Gold, Allan R. (1989) "Bush Administration Is Divided Over Move To Halt Global Warming," *The New York Times*, 27 October.

Goldenberg, Suzanne (2008) "Obama Breaks with Bush Oil Bosses and puts Environment at Top of Agenda," *The Guardian*, 16 December.

———— (2009a) "2500 Lobbyists, $45m On PR—But Just 12 Views Count," *The Guardian*, 13 May.

———— (2009b) "America's Gas Guzzlers On Road to Extinction As Obama Sets Strict Exhaust Emission Limits," *The Guardian*, 20 May.

———— (2009c) "Obama Ready to Use His Personal Popularity to Force Through Sweeping Climate Change Bill," *The Guardian*, 3 June.

———— (2009d) "Obama Pleads with Congress to Pass Historic Climate Bill," *The Guardian*, 26 June 2009.

———— (2009e) "Obama Offers Copenhagen Little Hope," *The Guardian*, 18 December.

———— (2014) "Obama Takes Historic Step to Reduce Carbon Emissions from Power Stations," *The Guardian*, 3 June.

Gore, Al (1992) *Earth in the Balance* (Boston: Houghton Mifflin). Revised edition published in 2013 (Abingdon: Routledge).

———— (1998) "Statement By Vice-President Gore on the United States' Signing of the Kyoto Protocol," 12 November.

———— (2011) "Al Gore: Climate of Denial," *Rolling Stone*, 22 June.

Greene, David L. (2010) "Measuring Energy Security: Can the United States Achieve Energy Independence?" *Energy Policy*, 38: 1614–21.

Grossman, Peter Z. (2013) *U.S. Energy Policy and the Pursuit of Failure* (Cambridge: Cambridge University Press).

Hale, Stephen (2010) "The New Politics of Climate Change: Why We Are Failing and How We Will Succeed," *Environmental Politics*, 19: 255–75.

Harris, Paul G. (2009) "Beyond Bush: Environmental Politics and Prospects for US Climate Policy," *Energy Policy*, 37: 966–71.

Harrison, David, Andrew Foss, Per Klevnas, and Daniel Radov "Economic Policy Instruments for Reducing Greenhouse Gas Emissions" in John S. Dryzek,

Richard B. Norgaard, and David Schlosberg (eds) *The Oxford Handbook of Climate Change and Society* (Oxford: Oxford University Press), 536–49.

Hart, David M. and David G. Victor (1993) "Scientific Elites and the Making of US Policy for Climate Change Research," *Social Studies of Science*, 23: 643–80.

Harvey, David (2005) *A Brief History of Neoliberalism* (Oxford: Oxford University Press).

HCC (House Committee on Commerce) (1995) "International Global Climate Change Negotiations." Hearing before the House Committee on Commerce, Subcommittee on Health and the Environment, 104th Congress, 1st session.

——— (1998) "The Kyoto Protocol and Its Economic Implications," Hearing before the House Committee on Commerce, Subcommittee on Energy and Power, 105th Congress, 2nd session.

HCoA (House Committee on Appropriations) (1956) "Second Supplemental Appropriation Bill, 1956." Hearing before US House of Representatives Committee on Appropriations, 84th Congress, 2nd session.

HCS (House Committee on Science) (1995) "Scientific Integrity and Public Trust: The Science Behind Federal Policies and Mandates: Case Study 2—Climate Models and Projections of Potential Impacts of Global Climate Change." Hearing before the House Committee on Science, Subcommittee on Energy and the Environment, 104th Congress, 1st session.

HCSST (House Committee on Science, Space and Technology) (1976) "The National Climate Program Act," Hearing before the House Committee on Science, Space and Technology, Subcommittee on Environment and the Atmosphere, 94th Congress, 2nd session.

Hecht, Alan and Dennis Tirpak (1995) "Framework Agreement On Climate Change: a Scientific and Policy History," *Climatic Change*, 29: 371–402.

Helm, Dieter (2012) *The Carbon Crunch: How We're Getting Climate Change Wrong—And How to Fix It* (Mew Haven, CN: Yale University Press).

Helm, Dieter and Cameron Hepburn (eds) (2009) *The Economics and Politics of Climate Change* (Oxford: Oxford University Press).

Hoffman, Andrew J. (2015) *How Culture Shapes the Climate Change Debate* (Palo Alto, CA: Stanford University Press).

Hopgood, Stephen (1998) *American Foreign Environmental Policy and the Power of the State* (Oxford: Oxford University Press).

Horner, Christopher C. (2008) *Red Hot Lies* (Washington, DC: Regnery).

Hulme, Mike (2009) *Why We Disagree About Climate Change* (Cambridge: Cambridge University Press).

——— (2014) *Can Science Fix Climate Change?* (Cambridge: Polity Press).

Hulse, Carl and David M. Herzenhorn (2010) "Democrats Call Off Climate Bill Effort," *The New York Times*, 22 July.

Inhofe, James (2003) "The Science of Climate Change." US Senate Floor Statement by Senator James M. Inhofe (R. OK), 28 July. Ihttp://inhofe.senate.gov/pressreleases/climate.htm. Accessed 24 October 2011.

IPCC (Intergovernmental Panel on Climate Change) (1990) *Climate Change: The IPCC Scientific Assessment* (Cambridge: Cambridge University Press).

———— (1996) *Climate Change 1995: The Science of Climate Change* (Cambridge: Cambridge University Press).

———— (2007) *Climate Change 2007: The Physical Science Base*, Contribution of Working Group 1 to the Fourth Assessment Report of the Intergovernmental Panel on Climate Change (Cambridge: Cambridge University Press).

———— (2013) *Climate Change 2013: The Physical Basis*, Working Group 1 Contribution to the Fifth Assessment Report of the Intergovernmental Panel on Climate Change (Cambridge: Cambridge University Press).

Jacques, Peter J, Riley E. Dunlap, and Mark Freeman (2008) "The Organisation of Denial: Conservative Think Tanks and Environmental Scepticism," *Environmental Politics*, 17: 349–85.

Jalonick, Mary Clare (2003) "Defeat of Senate Global Warming Bill Highlights Worries over Economic Impacts," *CQ Weekly*, 1 November.

———— (2004) "The George W. Bush Administration: A New Environment?," The French Center on the United States (CFE), May.

Jamieson, Dale (2011) "The Nature of the Problem" in John S. Dryzek, Richard B. Norgaard, and David Schlosberg (eds) *The Oxford Handbook of Climate Change and Society* (Oxford: Oxford University Press).

———— (2014) *Reason in a Dark Time* (Oxford: Oxford University Press).

Jehl, Douglas (2000) "The 43rd President: Interior Choice Sends a Signal on Land Policy," *The New York Times*, 30 December.

———— (2001a) "Environmental Groups Join in Opposing Choice for Interior Secretary," *The New York Times*, 12 January.

———— (2001b) "US Stance on Warming Puts Whitman in Tense Spot," *The New York Times*, 30 March.

Jehl, Douglas and Andrew C. Revkin (2001) "Bush in Reversal: Won't Seek Cut in Emissions on Carbon Dioxide," *The New York Times*, 14 March.

Jerit, Jennifer (2008) "Issue Framing and Engagement: Rhetorical Strategy in Public Policy Debates," *Political Behavior*, 30: 1–24.

Jones, Charles O. (1975) *Clean Air* (Pittsburgh, PA: University of Pittsburgh Press).

Jones, Jeffrey M. (2014) "In US, Most Do Not See Global Warming As A Serious Threat," Gallup Poll, 13 March.

Jordan, Andrew, David Benson, Rudiger Wurzel, and Anthony Zito (2011) "Policy Instruments in Practice" in John S. Dryzek, Richard B. Norgaard, and David Schlosberg (eds) *The Oxford Handbook of Climate Change and Society* (Oxford: Oxford University Press), 536–49.

Kahn, Greg (2003) "The Fate of the Kyoto Protocol under the Bush Administration," *Berkeley Journal of International Law*, 21: 3, 548–71.

Kaufman, Leslie (2010) "Darwin Foes Add Warming to Targets," *The New York Times*, 3 March.

Keller, Ann Campbell (2009) *Science in Environmental Policy* (Cambridge, MA: MIT Press).

Kingdon, John W. (2011) *Agendas, Alternatives, and Public Policies* (New York: Longman, revised 2nd edition).

Klyza, Christopher and David Souza (2008) *American Environmental Policy 1990–2006: Beyond Gridlock* (Cambridge, MA: MIT Press).

Knill, Christoph and Jale Tosun (2012) *Public Policy* (Basingstoke: Palgrave Macmillan).

Knox-Hayes, Janelle (2012) "Negotiating Climate Legislation: Policy Path Dependence and Coalition Stabilisation," *Regulation and Governance*, 6: 545–67.

Kolbert, Elizabeth (2006) *Field Notes from a Catastrophe* (London: Bloomsbury).

Kolk, And and David Levy (2001) "Winds of Change: Corporate Strategy, Climate Change, and Oil Multinationals," *European Management Journal*, 19: 501–9.

Krauss, Clifford and Jad Mouawad (2009) "Oil Industry Backs Protests of Emissions Bill," *The New York Times*, 19 August.

Kurtz, Howard (1997) "CNN Stops Airing Ad Campaign," *The Washington Post*, 3 October.

Kwa, Chunglin (2001) "The Rise and Fall of Weather Modification: Changes in American Attitudes Towards Technology, Nature and Society" in Clark A. Miller and Paul N. Edwards (eds) *Changing the Atmosphere* (Cambridge, MA: MIT Press).

Kymlicka, B.B. and Jean V. Matthews (eds) (1988) *Reagan Revolution* (Belmont, CA: Dorsey Press).

Landler, Mark (2014) "US and China Reach Climate Deal After Months of Talks," *The New York Times*, 11 November.

Landy, Marc (2010) "Adapting to Climate Change: Problems and Prospects," in Barry G. Rabe (ed.) *Greenhouse Governance* (Washington, DC: Brookings).

Layzer, Judith A. (2007) "Deep Freeze: How Business Has Shaped the Global Warming Debate in Congress" in Miachel E. Kraft and Sheldon Kamieniecki (eds) *Business and Environmental Policy* (Cambridge, MA: MIT Press).

Lazarus, Richard J. (2009) "Super Wicked Problems and Climate Change: Restraining the Present to Liberate the Future," *Cornell Law Review*, 94: 1153–234.

Lee, Gary (1993) "Clinton Offers Package to 'Halt Global Warming," *The Washington Post*, 20 October.

——— (1996) "U.S. Urges Binding Accord on Global Warming," *The Washington Post*, 18 July.

Lee, Henry, Vicki Arroyo Cochron, and Manik Roy (2001) "US domestic climate change policy," *Climate Policy*, 1: 381–95.

Leiserowitz, Anthony and Edward W. Mailbach (2009) "Global Warming and the 2008 Presidential Election," Yale Project on Climate Change Communication, 23 March. www.environment.yale.edu/climate/publication/global-warming-and-the-2008-election. Accessed 21 March 2012.

Lerer, Lisa 2009) "In Case You Missed It … Dems to W.H.: Drop cap-and-trade," *Politico*, 27 December.

Levin, Kelly, Benjamin Cashore, Steven Bernstein, and Graeme Auld (2012) "Overcoming the Tragedy of Super Wicked Problems: Constraining Our Future Selves to Ameliorate Global Climate Change," *Policy Science*, 45: 123–52.

Lewis, Paul (1995) "U.S. Industries Oppose Emission Proposals," *The New York Times*, 22 August.

Liffey, Kevin (1995) "U.N. Climate Deal Clinched, But What Does It Mean?" *Reuter European Union Report*, 7 April.

Lippman, Thomas W. (1993) "Energy Tax Proposal Has 'Green' Tint: Environmentalists Back Plan They Helped to Draft," *The Washington Post*, 2 March.

Lisowoski, Michael (2002) "Playing the Two-Level Game: US President George Bush's Decision to Repudiate the Kyoto Protocol," *Environmental Politics*, 11(4): 101–19.

Little, Alison (2003) "The Climate Bill Lost Out, the Environment May Yet Prove a Winner," *Grist*, 5 November.

Liu, Xinsheng, Eric Lindquist, and Arnold Vedlitz (2011) "Explaining Media and Congressional Attention to Global Climate Change, 1965–2005: An Empirical Test of Agenda-Setting Theory," *Political Research Quarterly*, 64: 405–19.

Lomberg, Bjorn (2007) *Cool It* (London: Marshall Cavendish).

Lutsey, Nicholas and Daniel Sperling (2008) "America's Botton-Up Climate Change Mitigation Policy," *Energy Policy*, 36: 673–85.

Lutzenhiser, L. (2001) "The Contours of US Climate Non-Policy," *Society and Natural Resources*, 14: 511–23.

Lyons, Michael (1999) "Political Self-Interest and US Environmental Policy" *National Resources Journal*, 39: 271–94.

McCarthy, Colman (1992) "Gore's Politics Are Ever Green," *The Washington Post*, 4 August.

McCright, Aaron and Riley E. Dunlap (2000) "Challenging Global Warming as a Social Problem: An Analysis of the Conservative Movement's Counter-Claims," *Social Problems*, 47: 499–522.

——— (2003) "Defeating Kyoto: The Conservative Movement's Impact on U.S. Climate Change Policy," *Social Problems*, 50: 348–73.

——— (2010) "Anti-Reflexivity: The American Conservative Movement's Success in Undermining Climate Science and Policy," *Theory, Culture, and Society*, 27(2): 1–34.

——— (2011) "The Politicization of Climate Change and Polarization in the American Public's Views of Global Warming, 2001–2010," *The Sociological Quarterly*, 52: 155–94.

Maniates, Michael and John M. Meyer (eds) *The Environmental Politics of Sacrifice* (Cambridge, MA: MIT Press).

Mann, Michael E. (2012) *The Hockey Stick and the Climate Wars* (New York: Columbia University Press).

Martel, Jonathan S. and Kerri L. Stelcen (2007) "Clean Air Regulation" in Michael Gerrard (ed.) *Global Climate Change and US Law* (Chicago: American Bar Association).

Matsuo, Naoki (2002) "Analysis of the U.S.'s New Climate Initiatives: The Attitude of the Bush Administration towards Climate Change," *International Review for Environmental Strategies*, 3(1): 177–87.

Mayhew, David S. (1991) *Divided We Govern* (New Haven: Yale University Press).

Meckling (2011) *Carbon Coalitions: Business, Climate Politics and the Rise of Emissions Trading* (Cambridge, MA: MIT Press).

Mencimer, Stephanie (2002) "Weather tis Nobler in The Mind," *Washington Monthly*, July/August.

Mendelsohn, Robert (2011) "Economic Estimates of the Damages Caused by Climate Change" in John S. Dryzek, Richard B. Norgaard, and David Schlosberg (eds) *The Oxford Handbook of Climate Change and Society* (Oxford: Oxford University Press).

Mirowski, Philip (2014) *Never Let a Serious Crisis Go to Waste* (London: Verso Books).

Monbiot, George (2007) *Heat* (London: Penguin Books).

Montford, A.W. (2011) *The Hockey Stick Illusion* (London: Stacey International).

——— (2012) *Hiding the Decline* (London: Anglosphere Books).

Mooney, Christopher (2006) *The Republican War on Science* (New York: Basic Books).

Morgan, Dan (1996) "Strenthened U.S. Commitment Lights a Fire Under Global Warming Debate," *The Washington Post*, 13 September.

Morrisey, Wayne A. (2001) "Global Climate Change: A survey of Scientific Research and Policy Reports" in Horace M. Karling (ed.) *Global Climate Change* (Huntington, NY: Nova Science Publisher).

Moser, Susanne C. and Lisa Dilling (2011) "Communicating Climate Change" in John S.Dryzek, Richard B.Norgaard, and David Sclosberg (eds) *The Oxford Handbook of Climate Change and Society* (Oxford: Oxford University Press).

Nader, Ralph (2002) *Crashing the Party* (New York: St. Martin's Press).

National Academy of Sciences (2001) "Climate Change: An Analysis of Some Key Questions," Committee on the Science of Climate Change, National Research Council.

Newport, Frank (2014) "Americans Show Low Levels of Concern on Global Warming," Gallup Poll, 4 April.

Nisbet, Matthew C. (2009) "Communicating climate change: why frames matter for public engagement," *Environment*, 51: 12–23.

——— (2011) "Public Opinion and Participation" in John S.Dryzek, Richard B.Norgaard, and David Sclosberg (eds) *The Oxford Handbook of Climate Change and Society* (Oxford: Oxford University Press).

Nordhaus, Ted and Michael Shellenberger (2007) *Breakthrough* (New York: Houghton Mifflin Harcourt).

NRC (National Research Council) (2011) *America's Climate Choices* (Washington, DC: The National Academies Press).

NRDC (National Resources Defense Council) (2001) "NRDC Says New Bush Global Warming Initiatives Are 'All Talk, No Action'," National Resources Defense Council, Press Release, 13 July.

Obama, Barack (2009a) *Memorandum on the Energy Independence and Security Act 2007*. 26 January.

——— (2009b) *Memorandum on the State of California Request for Waiver Under 42 U.S.C 7543(b), the Clean Air Act*. 26 January.

——— (2010a) *Memorandum Regarding Fuel Efficiency Standards*. 21 May.

Oliver, Mark (2005) "US: Climate deal complements Kyoto," *The Guardian*, 28 July.

Oreskes, Naomi (2004) "The Scientific Consensus on Climate Change," *Science*, 306: 1686.

——— (2011) *Climate Change Denial* (Abingdon: Earthscan).

Oreskes, Naomi and Erik M. Conway (2010) "The Denial of Global Warming" in Naomi Oreskes and Erik M. Conway *Merchants of Doubt* (London: Bloomsbury).

Payne, Rodger A. and Sean Payne (2014) "The Politics of Climate Change: Will the US Act to Prevent Calamity" in Ralph G. Carter (ed.) *Contemporary Cases in U.S. Foreign Policy* (Thousand Oaks, CA: CQ Press).

Peters, B. Guy and Brian W. Hogwood (1985) "In Search of the Issue-Attention Cycle," *Journal of Politics*, 47: 238–53.

Peters, Ronald M. and Cindy Simone Rosenthal (2010) *Speaker Nancy Pelosi and the New American Politics* (New York: Oxford University Press).

Peterson, Thomas and Adam Rose (2006) "Reducing Conflicts between Climate Policy and Energy Policy in the US: The Important Role of the States," *Energy Policy*, 34: 619–31.

Pew (2001) "Bush's Base Backs Him to the Hilt," The Pew Research Center, 26 April. www.people-press.org/2001/04/26/other-important-findings-and-analyses-12/. Accessed 26 June 2013.

——— (2007) "Global Warming: A Divide on Causes and Solutions," Pew Research Center, 24 January. www.pewresearch.org/pubs/282/global-warming-a-divide-on. Accessed 9 March 2012.

——— (2008) "Global Warming Falls Still Further on Republicans' Policy Agenda," Pew Research Center, 30 January. www.pewresearch.org/pubs/710/global-warming-republicans. Accessed 9 March 2012.

——— (2009a) "Economy, Jobs Trump All Other Policy Priorities In 2009," Pew Research Center, 22 January. www.people-press.org/report/485/economy-top-policy-priority. Accessed 13 July 2009.

——— (2009b) "Fewer Americans See Solid Evidence of Global Warming," The Pew Research Center for the People and the Press, October.

Pielke Jr., Roger A. (1995) "Usable Information for Policy: An Appraisal of the US Gobal Change Research Program," *Policy Sciences*, 28: 39–77.

———— (2000a) "Policy history of the US Global Change Research Program: Part 1. Administrative Development," *Global Environmental Change*, 10: 9–25.

———— (2000b) "Policy history of the US Global Change Research Program: Part 2. Legislative Process," *Global Environmental Change*, 10: 133–44.

———— (2011) *The Climate Fix* (New York: Basic Books).

———— (2014) *The Rightful Place of Science: Disasters and Climate Change* (Tempe, AZ: Consortium for Science, Policy and Outcomes).

Plimer, Ian (2009) *Heaven and Earth: Global Warming—The Missing Science* (Lanham, MD: Taylor Trade Publishing).

Pooley, Eric (2010) *The Climate War* (New York: Hyperion).

Posner, Paul L. (2010) "The Politics of Vertical Diffusion: The States and Climate Change", in Barry G. Rabe (ed.) *Greenhouse Governance* (Washington, DC: Brookings Institution).

Pralle, Sarah B. (2009) "Agenda-setting and climate change," *Environmental Politics*, 18: 781–99.

Putnam, Robert D. (1988) "Diplomacy and Domestic Politics: The Logic of Two-Level Games," *International Organization*, 42:427–60.

Rabe, Barry G. (2004) *Statehouse and Greenhouse* (Washington, DC: Brookings Institution).

———— (2007) "Beyond Kyoto: Climate Change Policy in Multilevel Governance Systems," *Governance*, 20: 423–44.

———— (2010a) "The 'Impossible Dream' of Carbon Taxes: Is the 'Best Answer' and Political Non-Starter?" in Barry G. Rabe (ed.) *Greenhouse Governance* (Washington, DC: Brookings).

———— (2010b) "Can Congress Govern the Climate?" in Barry G. Rabe (ed.) *Greenhouse Governance* (Washington, DC: Brookings).

———— (2010c) "Introduction: The Challenges of US Climate Governance" in Barry G. Rabe (ed.) *Greenhouse Governance*, (Washington, DC: Brookings).

Rahm, Dianne (2009) *Climate Change Policy in the United States* (Jefferson, NC: McFarland & Company).

Rajan, Sudhir Chella (1996) *The Enigma of Automobility* (Pittsburg, PA: University of Pittsburgh Press).

Ray, Julie and Anita Pugliese (2011) "Worldwide, Blame for Climate Change Falls on Humans," Gallup Poll, 22 April.

Revkin, Andrew C. (2002a) "US Sees Problems in Climate Change," *The New York Times*, 3 June.

———— (2002b) "With White House Approval, EPA Pollution Report Omits Global Warming Section," *The New York Times*, 15 September.

———— (2005) "Editor of Climate Reports Resigns," *The New York Times*, 10 June.

Revkin, Andrew C. and Katharine Q. Seelye (2003) "Report by EPA Leaves Out Data on Climate Change," *The New York Times*, 19 June.

Riker, William H. (1996) *The Strategy of Rhetoric* (New Haven, CN: Yale University Press).

Rittel, Horst and Melvin Webber (1973) "Dilemmas in a General Theory of Planning," *Policy Sciences*, 4: 155–69.

Rochefort, David A. and Roger W. Cobb (1994) "Problem Definition: An Emerging Perspective" in David A. Rochefort and Roger W. Cobb (eds) *The Politics of Problem Definition* (Lawrence, KS: University Press of Kansas).

Roper (1996) "Presidential Election 1996; Most Important Problem." http://www.ropercenter.uconn.edu/polls/us-elections/presidential-elections/1996-presidential-election/. Accessed 3 June 2015.

Rosenbaum, David E. (1992) "The 1992 Campaign: Running Mates; On Environment Issue, Quayle Curbed the Rules and Gore Wrote the Book," *The New York Times*, 11 July.

Rosenthal, Elisabeth (2009) "Biggest Obstacle to Global Climate Deal May Be How to Pay for IT," *The New York Times*, 15 October.

Royden, Amy (2002) "U.S. Climate Change Policy Under President Clinton: A Look Back," *Golden Gate University Law Review*, 32: 415–78.

Saad, Lydia (2014) "A Steady 57% in US Blame Humans for Global Warming," Gallup Organization, 18 March.

Samuelsohn, Darren (2009a) "Senators Spend Recess Fine-Tuning Messages on Cap and Trade," *The New York Times*, 28 August.

—— (2009b) "Boxer, Kerry Set to Introduce Climate Bill in Senate," *The New York Times*, 28 September.

—— (2009c) "Obama Negotiates 'Copenhagen Accord' With Senate Climate Fight in Mind," *The New York Times*, 21 December.

—— (2009d) "Has John McCain Gone Cool on Climate Change?" *The New York Times*, 16 July.

Sanger, David E. (2001) "Bush Will Continue to Oppose Kyoto Pact on Globalization," *The New York Times*, 12 June.

SCENR (Senate Committee on Energy and Natural Resources) (1988) "Greenhouse Effect and Global Climate Change, Part 2." Hearing before US Senate Committee on Energy and Natural Resources, 100th Congress, 2nd session.

—— (1994) "Energy Policy Act and the President's climate change action plan." Hearing before the Senate Committee on Energy and Natural Resources, 103rd Congress, 2nd session.

SCEPW (Senate Committee on Environment and Public Works) (1985) "Global Warming." Hearing before US Senate Committee on Environment and Public Works, Subcommittee on Toxic Substances and Environmental Oversight, 99th Congress, 1st session.

—— (1986a) "Global Warming." Hearing before US Senate Committee on Environment and Public Works, Subcommittee on Toxic Substances and Environmental Oversight, 99th Congress, 2nd session.

—— (1986b) "Ozone Depletion, the Greenhouse Effect, and Climate Change." Hearing before US Senate Committee on Environment and Public Works, 99th Congress, 2nd session.

———— (2009a) "Boxer Releases Chairman's Mark of Clean Energy Jobs and American Power Act," US Senate, Committee on Environment and Public Works, Press Release, 23 October. http://epw.senate.gov/public/index.cfm?FuseAction=Majority.PressReleases&ContentRecord_id=84691b8e-802a-23ad-4728-e60de8d50fea. Accessed 23 May 2012.

———— (2009b) "Boxer Statement on Committee Passage of S. 1733—The Clean Energy Jobs and American Power Act," US Senate, Committee on Environment and Public Works, Press Release, 5 November. http://www.epw.senate.gov/public/index.cfm?FuseAction=Majority.PressReleases&ContentRecord_id=c512ac4d-802a-23ad-4884-2b95a8405efe&Region_id=&Issue_id=. Accessed 6 November 2009.

———— (2009c) "Boxer Violates Committee Rules, Rejects Clear Path Forward," US Senate, Committee on Environment and Public Works, Press Release, 5 November. http://www.epw.senate.gov/public/index.cfm?FuseAction=Minority.PressReleases&ContentRecord_id=c4ec933f-802a-23ad-4a8c-ce4fb68c9650&Region_id=&Issue_id=. Accessed 6 November 2009.

Schneider, Keith (1993) "Gore Meets Resistance in Effort for Steps on Global Warming," *The New York Times*, 19 April.

Seelye, Katharine Q. (2002) "President Distances Himself from Global Warming Report" *The New York Times*, 5 June.

Selin, Henrik and Stacy D. VanDeveer (2007) "Political Science and Prediction: What's Next for US Climate Change Policy? *Review of Policy Research*, 24: 1–27.

———— (2009) *Changing Climates in North American Politics* (Cambridge, MA: The MIT Press).

Shabecoff, Philip (1988) "Global Warming Has Begun, Expert Tells Senate," *The New York Times*, 24 June.

———— (1989a) "Baker Urges International Action To Halt Global Warming Threat," *The New York Times*, 31 January.

———— (1989b) "EPA Proposes Rule to Curb Warming," *The New York Times*, 14 March.

———— (1989c) "Scientist Says Budget Office Altered His Testimony," *The New York Times*, 8 May.

———— (1990) "Bush Denies Putting Off Action on Averting Global Climate Shift," *The New York Times*, 19 April.

Shafire, David M. (2014) *Presidential Administration and the Environment* (New York: Routledge).

Shanley, Robert A. (1992) *Presidential Influence and Environmental Policy* (Westport, CN: Greenwood Press).

Shapiro, Sidney A. (2009) "'Political' Science: Regulatory Science After the Bush Administration," *Duke Journal of Constitutional Law and Public Policy*, 4: 31–44.

Shellenberger, Michael and Ted Nordhaus (2005) "The Death of Environmentalism: Global Warming in a Post-Environmental World," *Grist*, 14 January.

Shulman, Seth (2008) *Undermining Science: Suppression and Distortion in the Bush Administration* (Berkeley, University of California Press).

Simons, Marlise (1990) "US View Prevails At Climate Parley," *The New York Times*, 8 November.

Singer, Fred C. and Denis T. Avery (2007) *Unstoppable Climate Change Every 1500 Years* (Lanham, MD: Rowman and Littlefield).

Singer, Fred C, Roger Revelle, Chauncey Starr (1991) "What to do about Greenhouse Warming: Look Before You Leap," *Cosmos*, 1: 28–33.

Smith, Zachary A. (1992) *The Environmental Policy Paradox* (Englewood Cliffs, NJ: Prentice Hall).

Stagliano, Vito (2001) *The Policy of Discontent: The Making of a National Energy Strategy* (Tulsa, OK: PennWell).

Stainforth, Daniel A. (2005) "Uncertainty in Predictions of the Climate Response to Rising Levels of Greenhouse Gases," *Nature*, 433: 403–6.

Steffen, Will (2011) "A Truly Complex and Diabolical Policy Problem" in John S. Dryzek, Richard B. Norgaard, and David Schlosberg (eds) *The Oxford Handbook of Climate Change and Society* (Oxford: Oxford University Press).

Stevens, William (1997) "Gore, in Japan, Signals That US May Make Some Compromises," *The New York Times*, 8 December.

Stone, Deborah (1988) *Policy Paradox and Political Reason* (Glenview, IL: Scott, Foresman, and Company).

Sussman, Glen and Byron W. Daynes (2013) *US Politics and Climate Change* (Boulder, CO: Lynne Rienner).

Szasz, Andrew (2011) "Is Green Consumption Part of the Solution," in John S. Dryzek, Richard B. Norgaard, and David Schlosberg (eds) *The Oxford Handbook of Climate Change and Society* (Oxford: Oxford University Press).

Thaler, Richard and Cass R. Sunstein (2008) *Nudge* (Ann Arbor, MI: Caravan Books).

Torrance, Wendy E.F. (2006) "Science or Salience: Building An Agenda for Climate Change" in Ronald Bruce Mitchell (ed.) *Global Environmental Assessments* (Cambridge, MA: MIT Press).

Turner, Rachel (2008) *Neo-Liberal Ideology* (Edinburgh: Edinburgh University Press).

USDoD (US Department of Defense) (2014) "Quadrennial Defense Review 2014," US Department of Defense, Washington, DC.

USEPA (US Environmental Protection Agency) (1998) "EPA's Authority to Regulate Pollutants Emitted by Electric Power Generation Sources," Memorandum from Jonathan Z. Cannon, General Counsel, EPA, to Carol M. Browner, Administrator, EPA, 10 April, 4–5.

——— (2009) "Greenhouse Gases Threaten Public Health and the Environment," Press Release, EPA, 7 December.

——— (2010) "EPA and NHTSA Finalized Historic National Program to Reduce Greenhouse Gases and Improve Fuel Economy from Cars and Trucks,"

1 April. www.epa.gov/otag/climate/regulations/420f10014.htm. Accessed 30 March 2012.

——— (2011) *EPA and NHTSA Adopt First-Ever Program to Reduce Greenhouse Gas Emissions and Improve Fuel Efficiency of Medium- and Heavy-Duty Vehicles*, EPA, Office of Transportation and Air Quality, August. www.epa.gov/otag/climate/documents/420f11031.pdf. Accessed 30 March 2012.

——— (2012) "EPA Proposes First Carbon Pollution Standard for Future Power Plants/Achievable Standard is in Line with Investments Already Being Made and Will Inform the Building of New Plants Moving Forward," 27 March. http://yosemite.epa.gov/opa/admpress.nsf/bd4379a92ceceeac85 25735900400c27/9b4e8033d7e641d9852579ce005ae957!OpenDocument. Accessed 9 April 2012.

——— (2014) "Inventory of US Greenhouse Gas Emissions and Sinks: 1990–2011," www.epa.gov. Accessed 2 April 2014.

USGCRP (US Global Change Research Program) (2009) "Global Climate Change Impacts in the United States: 2009 Report," US Global Change Research Program, Washington, DC.

Vahrenholt, Fritz (2013) *The Neglected Sun* (London: Stacey International).

Vastag, Brian (2011) "Congress Nixes National Climate Service," *The Washington Post*, 20 November.

Vezirgiannidou, Sevasti-Eleni (2013) "Climate and Energy Policy in the United States: The Battle of Ideas," *Environmental Politics*, 22: 593–609.

Victor, David G. (2004) *The Collapse of the Kyoto Protocol and the Struggle to Slow Global Warming* (Princeton: Princeton University Press).

——— (2009) "On the Regulation of Geoengineering" in Dieter Helm and Cameron Hepburn (eds) *The Economics and Politics of Climate Change* (Oxford: Oxford University Press).

Vidal, John (2008) "Obama victory signals rebirth of US environmental policy," *The Guardian*, 5 November.

Vig, Norman J. (1994) "Presidential Leadership and the Environment: From Reagan and Bush to Clinton" in Norman J. Vig and Michael E. Kraft (eds) *Environmental Policy in the 1990s*, 2nd edition (Washington, DC: Congressional Quarterly Press).

Vogel, David (2012) *The Politics of Precaution* (Princeton, NJ: Princetin University Press).

van Vuuren, Detlaf, Michel den Elzen, Marcel Berk, and Andre de Moor (2002) "An Evaluation of the Level of Ambition and Implications of the Bush Climate Change Initiative," *Climate Policy*, 2(4): 293–301.

Walsh, Bryan (2011) "Is Obama Bad for the Environment?" *Time*, 6 September.

Watson, Harlan L. (2002a) "Press Briefing at the Eighth Session of the Conference of the Parties (COP-8) to the UN Framework Convention on Climate Change," New Delhi, India, 24 October. www.2001-2009.state.gov/g/oes/rls/rm/2002/14760.html. Accessed 31 March 2015.

———— (2002b) "Remarks to the Eighth Session of the Conference of the Parties (COP-8) to the UN Framework Convention on Climate Change," New Delhi, India, 25 October. www.2001-2009.state.gov/g/oes/rls/rm/2002/14758.html. Accessed 31 March 2015.

———— (2005) "US Climate Change Policy Overview," Seminar of Government Experts, Bonn, Germany, 16 May. www.unfccc.int/files/meetings/seminar/application/pdf/sem_pre_usa.pdf. Accessed 23 August 2013.

Weart, Spencer R. (2008) *The Discovery of Global Warming* (Cambridge, MA: Harvard University Press).

Weisskopf, Michael (1991) "Strict Energy-Saving Urged to Combat Global Warming," *The Washington Post*, 11 April.

Whitman, Christine T. (2005) *It's My Party Too: The Battle for the Heart of the GOP and the Future of America* (Toronto: Penguin).

Wilson, Kris (1995) "Mass Media as Sources of Global Warming Knowledge," *Mass Communication Review*, 22: 75–89.

Wolinsky-Nahmias, Yael (2015) *Changing Climate Politics* (Washington, DC: CQ Press).

Index

Printed and bound by CPI Group (UK) Ltd, Croydon, CR0 4YY

24/10/2024

01778283-0002